D0594355

A Manager's Guide to
INDUSTRIAL ROBOTS

A Manager's Guide to

INDUSTRIAL ROBOTS

Ken Susnjara

CORINTHIAN PRESS

Copyright © 1982 Kenneth J. Susnjara
All rights reserved. No part of this publication may be reproduced or transmitted in any form or by any means, electronic or mechanical, including photocopy, recording or any information storage and retrieval system, without written permission from the author.

Published by Corinthian Press
Publishing Division of EDR Corporation, Shaker Heights, Ohio

Printed in the United States of America

Library of Congress Catalog Card No. 81-86624
ISBN 0-86551-018-0

To my wife and children...
For the many evenings spent away from them preparing this work

INTRODUCTION

INTRODUCTION

It is difficult to pick up a periodical or trade journal today that does not contain some reference to industrial robots. In almost every area of manufacturing, robots are being considered for one or more tasks. Industrial robots have appeared on television and have appeared in newspapers and business publications. A good deal of excitement and expectation surrounds the area of industrial robotics today.

The robot industry is one of the fastest growing industries. The capabilities of these machines, the number of units available, the number of manufacturers, and the number of installations are all growing at an astounding pace.

A new industry growing like this is a two-edged sword. While the growth and proliferation of robots provide a great deal of opportunity, they also cause some large scale problems. The manager, whether a manufacturing manager, engineering manager, production supervisor, or the like, is caught in the middle. Upper management sees the robot as a way of substituting magic, dependable machines for difficult to manage personnel. They are sending edicts through the organization to implement the use of industrial robots. Production employees may view the industrial robot as a threat. Since it is a new field of endeavor, it is likely that your engineering staff does not understand industrial robots. And among all of this chaos it is a manager who is responsible for implementing the plans and making things happen, and this manager may not even know what an industrial robot looks like.

Not understanding robots or their implementation is not something to be ashamed of. Robots are a new phenomenon to such a large segment of industry that it is not reasonable to expect managers, especially nontechnical managers, to have a working knowledge of their use.

Having been in that position once myself, I found that there are some very fine publications covering the engineering aspects of industrial robots. The engineering detail of their application and installation is addressed by a number of competent authors, and anyone with an engineering background that is willing to put forth the effort can become familiar with industrial robots.

If you are a manager, however, especially a nontechnical manager, the problem becomes almost insurmountable. The engineering detail is either unintelligible or uninspiring and provides virtually no guidance in handling the manager's problems dealing with robotics.

This book has been written to try to fill this need. Together with my colleagues, I have found some common management problems while installing automation for the first time in over 100 plants during the last few years. We have seen reactions to those installations that both did and did not work. We have seen both inspiring successes and dismal failures. We have found the common mistakes that cause failure and observed the general pattern that insures success. In this book I have attempted to formalize this information and present it to the manager who, for better or worse, must now live in the world of robotics.

An attempt has been made to provide a manual which is as nontechnical as possible. I have assumed no formal technical education or technical background on the part of the reader. Since robots employ technology throughout, however, it is impossible to present them without some reference to the science and engineering of their operation. I have tried to define the technical terms as I proceeded and have supplied a glossary of terms in the simplest possible language in the back of the book.

The industrial world is changing rapidly. Robots seem to be developing a real and permanent place in our factories. I hope this work becomes a survival manual for those managers caught between an ever increasing drive for improved productivity through robotics and a lack of understanding and skills in dealing with the world of robots. By avoiding the mistakes of others and learning from their successes, hopefully even nontechnical managers will regard robots as an opportunity and an asset.

TABLE OF CONTENTS

Chapter One

WHAT IS AN INDUSTRIAL ROBOT?

Robot 1a: a machine that looks like a human being and performs various complex acts (as walking or talking) of a human being; also: a similar but fictional machine whose lack of capacity for human emotions is often emphasized; 2: an automatic apparatus or device that performs functions ordinarily ascribed to human beings or operates with what appears to be almost human intelligence; 3: a mechanism guided by automatic controls.

A friend of mine in the investment banking community spent the better part of a day discussing industrial robots with nontechnical investors. All through this discussion a series of scale models of various industrial robots sat on a desk. As the meeting was breaking up and the conversation turned from industrial robots, one of the visitors picked up one of the scale models, remarked that it was an interesting looking machine tool, and asked what it was.

To the uninitiated the word "robot" conjures up visions of mechanical creatures performing almost human feats. It at least conjures up a creature, like R2D2 from the recent movie *Star Wars*, which stands upright and squeaks in an unintelligible electronic voice. While these fantasy images of mechanical workers in industry have undoubtedly helped popularize the concept of industrial robots, they are unfortunately (or fortunately) not correct.

In an attempt to eliminate some of the confusion surrounding industrial robots, the Robot Institute of America (RIA) decided that a common, agreed upon definition of industrial robots was necessary. Repeated attempts to devise a simple, understandable definition failed before the following, somewhat gingerly worded, definition was adopted:

> A robot is a reprogrammable multifunctional manipulator designed to move material, parts, tools, or specialized devices, through variable programmed motions for the performance of a variety of tasks.

Definition per R.I.A.

The frustrating part about trying to develop this definition is that by looking at a machine it is quite simple to determine whether or not the machine should be classified as an industrial robot. Attempts, however, to commit this intuitive knowledge to paper in the form of a simple definition were frustrating at best.

The definition finally agreed upon does not, unfortunately, prove very enlightening to the uninitiated. A nontechnical manager who has not been exposed to industrial robots will glean little understanding of their looks or capability or functions from a simple definition. In an attempt to properly understand the function of industrial robots in manufacturing businesses today, it's necessary to go back and try to capture that intuitive sense of which machines are really robots.

Robots in the Work Force

The very first fact which must be understood is that an industrial robot is simply a machine. It is not a highly intelligent, mechanical person but is instead a functional machine tool. Industrial robots are today, and will likely remain for sometime in the future, extremely limited in what they can do when compared to even the most unskilled person. Even though some industrial robots are powered by very complex and capable computers, they still perform a limited sequence of motions.

So again let me repeat: *an industrial robot is a machine tool.* However, comparing its performance to that of a person is a unique attribute of industrial robots and brings us to what I feel is the core of the intuitive definition of an industrial robot. Industrial robots are, by their very nature, direct replacements for human labor. Their job seems to be performing various tasks that normally would be performed by a person and performing them in essentially the same manner that a person would. These statements concerning the purpose of industrial robots conjure up deep, intense fear on the part of those who manufacture robots. They fear that any mention of robots, i.e., mechanical people, might result in a serious backlash by thousands or millions of workers who fear they will be replaced by these mechanical people. These fears might be well founded except for some realities that are not being considered.

The first fact that must be understood takes us back to our number one premise that an industrial robot is nothing more than a specialized machine tool. Because it is a machine, it requires someone to program it and set it up, someone to keep an eye on it while it is running, even if only indirectly, and someone to fix it when it breaks. Each of these jobs requires some level of skill and special training. However, because the robot is a limited machine, it will need to be placed in a highly regimented environment and perform a fairly simple, repetitive, unskilled task. Because the unthinking, untiring industrial robot can perform its tasks consistently and unchanged day after day, it will provide considerable savings over the unskilled labor it replaces. People do not function well in unthinking, repetitive, monotonous, and burdensome tasks. These are in general the only types of jobs industrial robots are capable of performing. It is not surprising then that industrial robots will do a superior job in many of these tasks. At the same time, however, installations of industrial robots open up a number of new, exciting, and challenging jobs where none existed before.

The blue collar workers within the manufacturing sector who will be affected by the introduction of industrial robots are an astute and intelligent group of people. They are quite capable of understanding both the changes and the opportunities that will result from the implementation of industrial robots. They can understand as well as management the true nature of industrial robots and can assess, in most cases quite accurately, the effect industrial robots will have on them individually. The attempt by some manufacturers to call industrial robots by another less threatening name to avoid a problem must be quite insulting to those at whom it is aimed.

If industrial robots effectively and economically can replace unskilled labor, what then is keeping them from putting all unskilled personnel out of work? The answer to this question is quite subtle. However, it can provide valuable insights into the kinds of companies and the kinds of applications in which robots will prevail.

Let's go back to the basic premise that an industrial robot is a machine. As a machine, even though it performs the same task as a person, it is handled quite differently by a business. The cost of a machine is capitalized and then depreciated over a number of years. That depreciation is a very real cost on the profit and loss statement. If the funds to purchase the machine were borrowed, a long-term liability was created which must be repaid. An employee, however, is handled differently. An employee's pay each payday is counted as an expense against the production; however, an employee's future salary is not considered a liability in any form. Should a business find that it no longer needs to produce as much product, it can lay off workers and quickly reduce its costs. With robots in place of those workers, costs continue.

Another way to look at this same concept would be to take an example of a small company which has a task that can be performed by either an unskilled person or an industrial robot. Assume production requirements increase so that 50 additional jobs are created. If those jobs are filled by employees and accounted for in the normal manner, there will be no effect on the company's long-term liabilities, and the soundness of the company's financial statement as determined by the various ratios used by the banking community will not be affected. If, however, 50 industrial robots were installed, a large, inflexible, long-term liability will have been created. The potential for additional profits through lower costs and more consistent output from the industrial robots may be negated by the major negative impact on the financial statements.

Why, then, even consider industrial robots? This question brings us to the basic core of the industrial robot market. Industrial robots should be considered for the same reason that other pieces of automated machinery are considered. When these machines fill a need within a company, are cost justifiable, and provide more return for the available capital than any other, they should be installed.

In the first part of this recommendation, the word "need" was used. The need to use industrial robots will likely increase, possibly at a very fast pace.

About 1960 the growth of the population in the United States peaked and then began to decline. The population as a whole obviously continues to increase, but that increase began to occur at a slower pace after 1960. By 1980 we were to experience some interesting changes in industry. During the 1970s the number of young people entering the work force each year was greater than the year before. With the end of the conflict in Southeast Asia, a large number of available workers returned to the work force, and for the first time women in large numbers also entered the work force. This meant that through the 1970s enough labor was available to increase the total goods and services produced within our country (the GNP) without having to resort to higher productivity methods. Enough labor was available to increase production by simply increasing the number of people producing. In fact, during the late 1970s lower productivity methods were used to the extent that companies substituted labor for capital equipment. The lagging productivity growth figures through the 1970s bear this out.

You can't really blame American industry for taking this route. It is by far the safest and most reasonable method for increasing profits. Increasing production by simply increasing functions within the plant that are already proven, debugged, and to some extent perfected, seems reasonable, especially when compared to developing new techniques and new technology and making major capital commitments to new forms of automation. A large portion of the automation installed throughout the '70s was being installed to take advantage of a potential for increasing profits. Automation was being installed and justified based on the savings available to the manufacturer when compared with other methods. It is interesting to note that in calculating savings and costs of new programs, seldom, if ever, does a company consider the cost of learning to function with the new equipment.

Between 1980 and the end of the century the number of people entering the work force each year will be less than the year before. This being the case, much of the thinking and many of the business techniques for expansion used throughout the '70s will no longer suffice in the '80s. Labor shortages could easily become commonplace. As shortages of special skills develop, manufacturers will be forced to provide training to their employees to develop these skills, moving the general level of competence within the plant to a higher level. The size of the unskilled labor force could easily shrink during this period, putting tremendous pressure on industry to automate. Part of this problem can be alleviated by a relaxation of immigration policies; however, great pressures will develop for the use of industrial robots.

Already today, monotonous, burdensome, repetitive jobs are becoming difficult to fill. Great pressures are being brought on manufacturing to remove people from hazardous spray booths, coating chambers, and so forth. Even unskilled laborers are becoming more selective in the type of jobs they are willing to do. The population trends already established cannot help but make this problem worse.

By reading the popular business periodicals and newspapers and examining comments from political leaders both on state and national levels, the beginning of a ground swell to increase productivity and reindustrialize the nation can be seen. This ground swell is a reaction to problems already being felt which will be a major driving force through the '80s. In light of the above, industrial robots will be a part of virtually every production facility in the country. They will not be replacing workers already on the job; instead, they will provide the future unskilled work force. A realization that this is occurring is commonplace today. This is further evidenced by the number of managers who feel it necessary to learn more about this new, emerging technology. This may be the reason you are reading this right now.

In establishing the concept of an industrial robot, we have already determined that an industrial robot is first and foremost a machine tool. What makes robots unique when compared to other machine tools is that (with apologies to those who fear this part of my definition) they are designed to perform tasks normally performed by a person. This, however, is not quite precise enough, since many pieces of fixed automation rapidly perform tasks that at one time might have been done by hand. As an example, an automatic machine that assembles a ball point pen certainly performs a task normally done by a person; however, few if any would consider such a machine an industrial robot. Therefore, it's necessary to add one additional concept to this definition. The tasks being performed need to be performed in essentially the same manner at essentially the same speed that a person would perform them.

An industrial robot is now being defined to be a piece of machinery that does a job in approximately the same manner as a person does. While it would be easy to point out industrial robots that can lift more, reach farther, or move faster than a person, this definition can provide, in a nontechnical and nonprecise manner, a good guideline as to what an industrial robot really is.

Engineering Principles

Now that you have a conceptual idea of what an industrial robot is and what it does, a common question of the uninitiated is, "What does it look like?"

Industrial robots come in a variety of sizes, shapes, configurations, and complexities. In order to understand this area it is necessary to segment robots into a number of different categories.

Before discussing these categories, it is necessary to understand the meaning of several engineering terms. The first term which must be understood is "axis." You will find that robots are many times specified by the number of degrees of freedom or axes contained. An axis is a degree of freedom or a basic motion allowed by the mechanism. As an example, the bedroom door in your home is considered by engineers as a one axis mechanism. The door is capable of pivoting around a line which goes through the center of the door hinges to which it is mounted. The door's swinging open or shut is the one degree of freedom available. It cannot, however, move up and down, tilt, or move in any other direction other than swinging back and forth around the hinge. Another one axis mechanism would be a child's electric train riding on a track. The only motion available to the train is either forward or backward along the track. Even though the track may curve or go straight at different points, the fact that the train is restricted to simply moving forward or backward and is guided (it cannot move side to side, up and down, etc.) makes it a single axis mechanism. From this you can see that an axis can be both linear like the electric train or rotary like the bedroom door, still restricted to a single degree of freedom. Each independent slide or rotary joint within a robot, then, is referred to as an axis. The electric train can serve as an example to explain the difference between a servo controlled and non-servo controlled axis.

If you had a spot on a long section of straight track at which you wanted to stop the toy train, there would be two different methods which could be used to accomplish this. The first and simplest method would be to find a very heavy obstacle like a concrete block, place it on the track in such a way that when the train is touching it, it is in the desired position. It is then necessary simply to run the train in the proper direction and wait for it to hit the block. Then it is properly located. This system, with a bit more finesse but not much more, is called a non-servo controlled system.

A second way to locate the train at the desired position would be to turn it on and as you watch it near the proper position, use the electric control to slow it down until it is properly located, then turn off the electric control. If you substitute a sensor that can tell how far the train is from the desired stopping point and allow the sensor to operate the electric control in place of the person, you have a servo controlled system. The major advantage of a servo controlled system is that it can stop at any point along the track without having to reposition the block.

Non-Servo Controlled Robots

As you might guess from this, robots are generally separated into either servo or non-servo controlled types. The non-servo controlled robots are generally fitted with mechanical stops and are driven into these end point stops, which define the desired positions. Stops between the end points can be cycled into place to provide intermediate positions other than the end points of the axes. It is, however, apparent that the major disadvantage of a non-servo type industrial robot is the limited number of points at which it can stop. This should not, however, be regarded as a condemnation of this type of machine. A tremendous number of tasks performed within industry can be efficiently and inexpensively handled by this type of simple robot.

Non-servo type robots are generally quite a bit less expensive than their corresponding servo type systems. Many are capable of using much simpler control systems than the electronic computer control found in most of the servo controlled machines. Simple air logic controls or electrical sequencing controls perform quite adequately with the non-servo type robot. These control systems can be obtained at a much lower cost than the servo type control systems.

Electrical sequencing controls come in many different configurations; however, all have one thing in common. They provide a program signal or signals *to* the robot and wait for a signal *from* the robot telling the control that some event has occurred. This event can be something as simple as an arm extension or a clamp closing. Once the signal indicating that the event has occurred is returned to the controller, the controller steps to the next preprogrammed combination of signals which is then sent *to* the robot. Again the controller waits for a signal indicating that the necessary event has occurred, at which time the controller steps to the next set of signals. In this way the programmer sequentially steps through a series of preprogrammed signals with each new step actuated by a signal from the robot indicating that the last step is complete.

Air logic control works very much like the electrical sequencing control except no electrical connections are necessary. All the steps and signals are controlled by the operation of a series of air valves. This type of air logic control has very definite advantages when operating in explosive atmospheres. Since no electrical signals are present, the chance of a spark igniting the environment is reduced.

While non-servo type robots are generally small and designed to handle small parts at high speed, non-servo machines are available which can handle parts weighing in excess of 100 pounds and moving over a fairly large area.

Non-servo type robots can provide surprisingly close accuracies at the end points of their travel. Since they generally operate against fixed, mechanical stops, the end accuracy of the robot is dependent on the mechanical give or stop that has developed in the system. This can be kept to a minimum. Small air operated non-servo robots can easily hold .001" overall accuracies, while .001" to .002" end point accuracies on some of the larger non-servo robots are possible.

Servo Controlled Robots

Servo controlled machines come in many sizes, shapes, and configurations. They are endowed with a variety of working envelopes and weight carrying capabilities. Much of the remainder of this book about industrial robots will direct itself toward these various servo controlled robots.

In general, servo controlled robots are much more capable than the non-servo type robots. They are also more expensive, although in recent times the most expensive non-servo robots and the least expensive servo robots are overlapping in price.

Servo controlled robots are generally controlled utilizing micro electronics and a computer controlled system. These controls provide the robots with a variety of different capabilities which are difficult or impossible to achieve using the non-servo type control. Computers have become a part of many products you use today. Some understanding of their operation is important and will become more important in the next few years. Computers started as large expensive systems, but with increased volume production they came down in price so that even the smallest business could afford them. They are now available for home use, and before long you will find them in most of your home appliances, television sets, automobiles, and the like. Within a few years computers will be a part of almost every aspect of daily life, and those who do not have a knowledge of their operation and basic functions may be considered functional illiterates.

Before you panic, let me assure you that even though computers may be very large and complex their basic operations and functions are extremely simple. The principles behind the operation of a computer are understandable by virtually anyone. You may take some solace in the fact that even those who work with a computer day in and day out, analyzing and programming, do not really understand all the intricacies of its operation.

A computer is essentially nothing but a very large collection of electronic switches. These switches can either be on or off, with one position representing the digit 1 and the other position representing the digit 0. The digits 1 and 0 are called "binary bits." A string of eight of these binary bits is normally called a "byte." These bytes then represent various numbers and letters. Within the computer there is a series of instructions such as "add," "subtract," "compare," and the like. These instructions perform various functions on the strings of binary bits. The computer program is nothing more than a sequence of these computer instructions.

In order to provide a useful function the computer must first accept data, compute the data, remember parts of the data that must be retained, and in some way communicate the answer to the operator. In order to perform these functions several additional devices are needed. First, an input device is necessary to feed in initial data. These devices are normally keyboards which look very much like a standard typewriter. As the information is typed in, it is displayed on a small television screen or "CRT." "CRT," by the way, stands for "cathode ray tube." In order to remember various parts of the program, several different types of memory devices are used today. The most common is a small chip called a "RAM" or "random access memory." These RAM chips can be given information which they will retain until they are told to erase it and replace the information with new information. So in essence, a RAM chip is nothing more than a simple scratch pad or blackboard within the computer on which the computer is capable of storing information as necessary. RAM chips, however, have one disadvantage. When power is removed or the computer unplugged, they go blank, or they erase themselves. This type of memory is called "volatile" memory. "Volatile" means that the memory will be lost whenever power is removed from the system. Should the data stored need to be maintained even if the power is turned off, then a type of "non-volatile" memory must be used. There are a number of different types of non-volatile memory in general use today. The first of these is a chip which looks very much like the RAM chip discussed before. This chip, however, is called a "ROM." "ROM" stands for "read only memory." These chips have the information or data permanently built into them, and this information can be read as many times as necessary; however, it cannot be erased and replaced with new data. The data within ROMs is permanent and is installed when the chips are first made.

A variation of the ROM is the "PROM." "PROM" stands for "programmable read only memory." These chips are designed so that data we wish to store can be read into the chips once, at which time it becomes permanent and can no longer be altered or erased. In many control systems the instructions used to operate the machine or operating system are stored in either ROM or PROM.

There are a number of magnetic type devices used to store computer data. The simplest of these is the magnetic tape. This tape, which is essentially the same kind used in audio tapes and 8-track tape players, is used extensively for backing up a system. "Backing up" means storing everything from the system on a reel of magnetic tape which is kept separate from the system. This is done in case something should happen to the computer or installation which might destroy the information present. A properly backed up system can be completely reconstructed from the back up tape. The disadvantage of tape, however, is that it takes a great deal of time, at least as far as the computer is concerned, to record and retrieve information.

In order to alleviate this situation, another product called a magnetic disc was developed. This disc is nothing more than a round, flat disc of magnetic material, the same type used in the audio tape, which spins at approximately 3,500 RPM. A pickup head rides above the spinning disc and can move in and out very quickly to record or retrieve information from the disc. Discs are used for storage of a large volume of information which must be accessed quickly. Just like the magnetic tapes, magnetic discs have non-volatile type memories and do not require a continual source of power.

The type of disc we have been describing is the hard disc type normally used in large business systems. These discs are 12" to 14" in diameter, are made of a rigid, hard material, and can be stacked 4 or 5 high. The discs have magnetic material on both the top and the bottom, and magnetic heads can read from both positions. New smaller versions of the hard disc are now available in 8" and 5-1/4" diameter sizes, providing relatively low-cost mass storage.

Another type of disc which is available is the floppy disc. A floppy disc works in essentially the same manner as a hard disc except that the magnetic material is a thin, flexible plastic disc approximately 8" in diameter. This disc is contained in a paper container, and the whole unit is quite flexible; hence the name floppy disc. These discs can be removed and replaced in the disc reader quite easily and quickly. This provides a very convenient mechanism for recording and storing computer data as a back up. Floppy discs do not store quite as much information as a hard disc, but they are also considerably less expensive. A floppy disc can have densities up to approximately one megabyte (one million bytes of information), while hard discs of 100 to 500 megabytes are available.

Output devices can take a variety of different forms. These devices are the method by which the computer communicates to the outside world. The most common output device is the CRT. (Remember the television tube I talked about.) On this tube the computer can print our various prompts, messages, and answers in communicating with the computer operator. A second common output device is the printer. This is a device for printing out various data, reports, forms, and the like. In a computer controlled robot installation the robot actuator itself is an output device. The output from the computer is the actual motions of the robot.

A typical computerized system consists of a CPU or central processing unit which is the actual calculating part of the computer, a section of computer RAM memory, an input device of some sort, a mass program storage such as a disc or a floppy disc, and an output device.

A last but important part of a computerized system is the program. As we stated earlier the program is simply a set of instructions to the computer telling it what to do. These instructions are nothing but a string of 1s and 0s. The program as depicted by these strings of 1s and 0s is said to be written in machine language. No one I am aware of today is capable of writing programs in machine language. The normal method of writing computer programs is to use what is known as a high level language. Many of the terms that you may have heard, such as fortran, cobol, pascell, basic, or algal, are simply the names of various types of high level languages. When using a high level language the instructions you wish to perform are written using simple words such as get, save, add, divide, and the like. The high level language is actually a program of its own written in machine language which converts the high level words to machine language. As I once read, high level languages are developed by geniuses for use by morons. By using high level languages, therefore, computer programming is not all that mysterious or difficult: Get the first number; get the second number; add them together; place them on the CRT screen. This simple string of commands might easily be a computer program in a high level language for adding two numbers together and displaying them on the screen. Thousands of people write computer programs every day. High school students and even grade school children can and do write computer programs today. So believe me, there is nothing about computers that is beyond the normal individual endowed with simple, basic intelligence.

In dealing with the control systems of industrial robots, you will have little or nothing to do with this type of computer programming. However, some knowledge of and background in how the process works is oftentimes useful.

The computer control system used on industrial robots is very close to the system used in administrative machines. The computer control generally has two types of memory. The operating system, that is, the programs that instruct the robot how to operate but not necessarily which actions to take, are contained in some type of non-volatile memory, generally either PROMs or ROMs. A section of memory is then utilized for storing the user programs, that is, the sequence of motions and actions you wish the robot to produce. This memory is generally RAM memory. Because this memory is volatile, most industrial robots have some form of backup so that once the program is developed it can be saved and reused in the future. This backup can either be in the form of magnetic tape or floppy disc.

The input device for most computer controlled robots is some type of teach pendent. This teach pendent is very much like the computer console used in administrative computers to communicate with the machine. This teach pendent in many cases is also an output device that instructs and helps the programmer with necessary steps in developing a program. The major output device, however, is the robot actuator itself. Various calculations and information that the computer control system of an industrial robot outputs are converted by the electronics in the control system to various movements, motions, and events on the actuator itself.

In addition to classifying robots by their control system, either servo or non-servo, there are several other classifications of industrial robots in general use today. One of the more common classifications groups industrial robots by their operating methods. In this classification robots are classified as either pick and place, point to point, or continuous path.

Pick and Place Robots

The pick and place designation is normally reserved for the non-servo type machine with fixed stops at the end of each axis. These machines normally have a limited capability. They are able to perform a limited sequence of events but at times accomplish these at very high speeds.

As I stated earlier, these simple pick and place machines can also operate at very high accuracies. The pick and place machines are generally of the polar coordinate system, and most are small compared to the larger point to point and continuous path machines. Many of these machines are air powered with very simple control systems.

I guess the name "pick and place" comes from the fact that in general the task this machine performs is that of moving to a position, grasping a part, removing it, moving to a second position, and inserting the part.

Some of the more sophisticated pick and place machines have intermediate stops that can be cycled into and out of position during the program. In this way each axis can be stopped at more than two positions. For example, by cycling stop number 1 into position and then operating the air cylinder or hydraulic cylinder driving the axis, the axis can be moved against stop 1, defining an end point position. By then retracting the cylinder, moving stop 1 out of position, and moving stop 2 into a different position, cycling the same air or hydraulic cylinder will move axis 1 against stop 2, defining a second position. In this manner fairly complex programs can be developed to pick a part from one position and place it in another.

The next two classifications, point to point and continuous path, are used to describe two different methods of operating a servo controlled robot. These classifications refer to the method whereby programs are input, stored in the computer, and automatically played back.

Point to Point Robots

The point to point servo controlled robot is capable of moving to any point within its working envelope. Programming a point to point robot is accomplished with some type of teaching terminal, pendent, hand held programmer, or the like. Using this device the robot is operated at slow speeds and is first moved to a point which is to become the first point in the program. When the point is achieved, an "enter" button records that point in the computer. Each of the axes are then operated again, using the teach pendent until the second point in the program is reached. Again the "enter" button is pressed recording the second point. It is important to note that the only information stored in the computer control system is the location of each axis of the robot when the "enter" button is pressed. The sequence by which the programmer is able to move the machine to the desired point is not retained in any way. A variety of programming aids are available, with each manufacturer claiming features that make its machine easy to program. These aids normally are designed to make the desired point easy to achieve using the hand held programmer. Some of these aids are more useful than others; however, the basic concept of moving the machine to each of the desired points in sequence and entering each is the common programming method for point to point robots.

In executing the program the robot will move to the first point in the program, then proceed in an approximate straight line to the next point, then to the next, and so forth until it has executed the entire program. It is necessary to remember in programming point to point robots that the robot will move in a straight line between the points and not necessarily along the path that the arm was moved during programming. This fact can cause some difficulty when an inexperienced programmer is learning to program an industrial robot.

There is one disadvantage to this type of programming: The robot must be taken out of production and made available during programming. For this reason the idea of remotely programming a point to point robot in the office, as is done with numerically controlled machine tools, seems to have some merit. Doing this, however, presents several major obstacles. The first is that generally the programmer only knows the point at which the end of the arm or the end of the tool should be placed and does not necessarily know the position of each axis of the robot needed to reach that point. When the robot is moved so that the end point is in the proper position, it then simply reads the position of each of the axes to determine how they must be configured in order to achieve the final end point. When the end point is programmed in an office away from the robot, another method of determining the various joint and slide positions must be developed. One way of accomplishing this is to have the computer mathematically calculate the positions of each of the joints on the robot needed to achieve the desired end point. This sounds simple, but, in fact, is very complex; however, it has been accomplished today by at least one manufacturer and several universities and robotic research groups.

Another problem which complicates the first is that the joint sensors are not necessarily the same on identical robots. A robot joint provides the computer with a signal when it is in a certain position. That robot always will present the same signal when it is in that position. In this way, recording the signal stores the position, and driving the joint until that signal is achieved produces the necessary position. Another robot, however, of the same make and model, using exactly the same sensor, may present a slightly different signal when the joint is in that same position. Because of this problem, programs developed on one machine may not exactly duplicate on a second machine; therefore, programs developed theoretically in the office will operate differently on different pieces of equipment.

To get around this problem a correction table method has been developed which provides consistent positioning of robots regardless of normal variations in the joint sensors. In this system all mechanical positions that a joint can take are developed while the signals of the joint sensor are monitored. For each mechanical position of the joint a signal from the sensor is read and compared to the standard signal desired from that mechanical point. If the sensor reading is different than the standard, a correction factor is stored in a special table so that the signal can be corrected each time it must be used. This correction table must be developed each time either the joint or the sensor is taken apart and will be different for each machine manufactured. This system does, however, allow a mechanical position of the robot to correspond to a precise and predictable sensor signal.

As these capabilities become practical and more common, the dreams of a completely automated factory driven by a CAD/CAM system seem more obtainable. A CAD/CAM system (computer aided design/computer aided manufacturing) is a system in which product to be manufactured is designed with the help of a computer which develops all of the necessary programs for the NC machine tools, manipulators, and robots, and with the aid of computers automatically manufactures the desired products. This type of system is the ultimate in automation, and while the first example of these may be seen in the next five to ten years, widespread use is still in the future.

Another capability which has been given to servo controlled point to point robots is line tracking and position transformations. Line tracking is very simple. It means that a program developed on a stationary or stopped part can be executed later while the part is moving down a conveyor line. In fact, the speed of the line may vary, and the conveyor may stop and even reverse direction while the robot automatically compensates for the line movement.

Line tracking is accomplished today in two ways. The first and simplest method is one in which the robot simply locks onto the line and moves physically with it while it performs its task. If the line speeds up, slows down, or reverses direction, so does the entire robot. Although not as sophisticated as other methods, this system is very simple, easy to understand, and practical.

A second type of line tracking is accomplished with a method called "mathematical transformations." In this system the robot mathematically changes the program in space, "transforming" the program to allow for conveyor movement. In order to accomplish this, very complex mathematical capabilities are necessary. This capability is available from at least two major manufacturers today and will likely become more common as time goes on.

Continuous Path Robots

In certain applications, such as spray painting, industrial robots are called upon not only to move to distinct points in space, but to move along predetermined paths. In performing a spray painting operation the robot must be able to duplicate accurately the curved, flowing motions of a person using a spray gun. Obviously, trying to accomplish this using a point to point robot would be difficult at best. In order to accomplish these kinds of tasks, a new type of robot has been developed. This robot is the continuous path robot.

Continuous path robots are designed so that their structure is somewhat lighter and less massive than that of the point to point robots. They are counterbalanced using springs or other methods, and the drive systems are such that they may be disengaged allowing a person to move the arm about physically. Programming a continuous path robot generally requires an individual to grasp the end of the arm and move the arm carrying the tool, i.e., a spray gun, through the desired path.

The continuous path robot operates internally very much like the point to point robot. Instead of relying on the operator, however, to indicate the various points to which the robot is to travel, a continuous path control system records the position of each axis many times per second, with each of those recordings being an individual point. In running the program the points are played back at the same rate they were recorded, and the robot arm attempts to move to each of these points as they are played back. The resulting motion is a very close approximation of the original path which was entered.

Because the continuous path robot records many, many more points than the point to point robot, the electronic memory required is considerably larger. For this reason, plus certain other characteristics of the continuous path robot, it is generally more expensive than the corresponding point to point robot.

Continuous path robots normally have less load carrying capability than the larger servo controlled point to point robots. Since the arms and actuators have been designed to be as lightweight as possible for programming, their ability to carry heavy loads is somewhat diminished.

Continuous path robots generally have the capability of moving at very high speeds. This is necessary so that adjustments required to perform the programs can be accomplished. The smooth flowing motions of a continuous path program make these machines look very humanlike in their actions and motions. This, of course, is because their motions are very humanlike; they are the motions input by the programmer.

Another classification system for industrial robots distinguishes the robot by the working envelope. I have mentioned working envelopes several times, and possibly a clarification of what a working envelope is might prove helpful.

A working envelope is simply the shape of the entire area that the end of the robot arm can reach. Anything within the working envelope can be reached by the end of the robot arm; anything outside of the working envelope cannot be reached by the end of the robot arm. The shape of this area or working envelope is determined almost exclusively by the mechanical configuration of the robot.

There are four basic work envelope shapes. These four shapes are used by almost all of the robots produced today. The four shapes are rectangular, cylindrical, spherical, and jointed arm. Diagrams of typical robots operating in each of these configurations are shown in Figures 1.1 through 1.4.

Rectangular Robots

Referring to Figure 1.1, you can see the general mechanics and work envelope shape of a rectangular coordinate robot. As you can see in this type of machine, all axes are linear or slide axes. Axis 1 is a slide left to right. Axis 2 is shown as a slide in and out, while axis 3 is a slide up and down. The shape of the work envelope as traced by

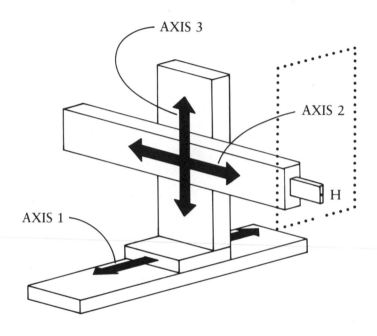

Figure 1.1
Rectangular Coordinate Robot

end point H can be seen by the dotted line. While the use of rectangular coordinates for industrial robots has not been common in the U.S., except for some arc welding robots, there are several European assembly type robots which use the rectangular coordinate system. In assembly work where precise points must be achieved and part assembly or inserting requires an up and down motion, these machines perform very well.

Cylindrical Robots

Referring to Figure 1.2, a general idea of the mechanics of a cylindrical coordinate robot can be seen. Axis 1 at the base allows the robot arm to rotate. Axis 2 moves the horizontal arm up and down, while axis 3 moves it in and out. A maximum point which can be achieved by the end point H traces the shape of a cylinder as shown by the

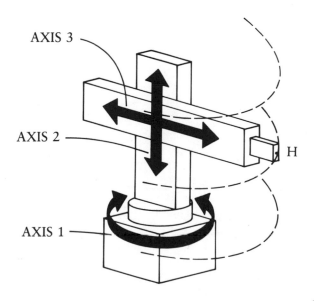

AXIS 3

AXIS 2

H

AXIS 1

Figure 1.2
Cylindrical Coordinate Robot

broken lines. Machines that operate in this manner are called cylindrical coordinate machines. Many if not most of the non-servo point to point machines use the cylindrical coordinate configuration. The long slides lend themselves well to the installation of physical stops or limit switches. There are, however, some sophisticated servo controlled industrial robots that utilize the cylindrical coordinate system.

Spherical Robots

Figure 1.3 shows a typical spherical coordinate robot. Axis 1 provides rotation on the base, while axis 2 allows the main body of the robot to rock point H up and down. Axis 3 allows the slide to move in and out. At its extremes point H traces the shape of a sphere; hence the name "spherical coordinate system." The largest number of spherical coordinate machines are servo controlled; however, there are some non-servo controlled spherical coordinate machines on the market.

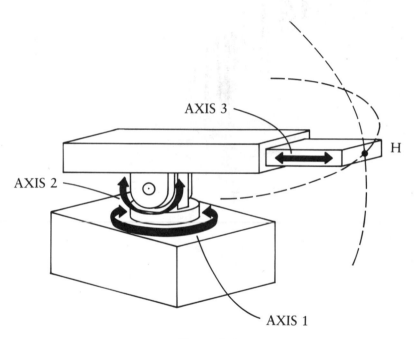

AXIS 3

H

AXIS 2

AXIS 1

Figure 1.3
Spherical Coordinate Robot

The spherical coordinate system was the one chosen by Unimate over twenty years ago for its line of industrial robots. This line of robots has been very successful; hence, a large number of spherical coordinate robots are in operation in industry today. The system has proven very effective, and these robots are being utilized today to perform a wide variety of tasks.

extremes causes point H to trace out an envelope shape shown in Figure 1.4. The jointed arm design in general provides the largest working envelope per area of floor space of any of the robot designs. The anthropomorphic design, however, requires a coordinated movement of each of the rotary axes in order for point H to move in a straight line between one point and another. This required coordination can only be accomplished utilizing a computer control system. Therefore, the jointed arm or anthropomorphic design is utilized today only by servo controlled type robots. Even with the use of computer systems, providing an anthropomorphic type robot with the ability to develop a straight line between two points in space requires rather sophisticated control capabilities.

In trying to determine which of the robot configurations is best, it is necessary to determine the requirements of the particular application under consideration. Each of the working envelope shapes and mechanical configurations has its particular advantages and disadvantages. It is necessary when attempting to place an industrial robot in a job to select those machines whose advantages can be utilized and whose disadvantages can be minimized. Using a machine with the proper mechanical configuration and working envelope can go a long way toward simplifying any particular application.

The definitions and classifications in this chapter are by necessity general, and it must be understood that many variations and permutations exist. Each robot manufacturer is forced to choose from the many options available in designing its product line. Once a general understanding of robotics design and operation has been developed, it will be necessary to compare available robots by utilizing each individual manufacturer's product literature. By examining various trade journals and other publications, various robot to robot evaluations have been and will continue to be made as an aid to prospective robot users.

You may have noticed by now that each of these designs a contains only three axes. Most industrial robots contain fiv independent axes; however, the two or three additional axes mally utilized where we show point H. These axes operate which provides articulation to the end of arm tooling or "h the robot.

Jointed Arm Robots

The fourth and possibly most complex working envelope anthropomorphic or jointed arm. This anthropomorphic or j arm design most closely resembles the workings of the human a is made up of a base rotation axis 1, a shoulder rotation axis an elbow rotation axis 3. The operation of these axes to

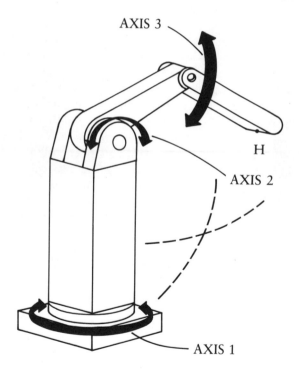

Figure 1.4
Jointed Arm Robot

Chapter Two
HOW DOES IT WORK?

With this background on industrial robots in mind, it is time to look at how an industrial robot actually works. I am not going to discuss the engineering aspects of an industrial robot as much as the practical aspects of actually programming and using a robot. In order to understand the things that are necessary to utilize an industrial robot, some basic understanding of how robots function is a must. I will, however, attempt in this discussion to keep from delving too deeply into the engineering details.

In order to understand the operation of a robot, it's helpful to divide the robot into three distinct parts. The first of these is the actuator or the actual, physical base and moving arm of the robot. Arms can be designed in several configurations as discussed in the previous chapter. These actuators can be powered by either electric motors, compressed air, or hydraulic fluid under pressure.

The second basic component of an industrial robot is the power source. This source may be electric motors in an all-electric industrial robot, an electric motor and hydraulic pump in a hydraulic powered industrial robot, or compressed air in a pneumatically powered machine.

The third basic component is the control system. Control systems also come in many configurations, from simple cams and rotating drums to sophisticated computerized electronic controls.

These three basic components can be configured in many ways depending on the end application and the manufacturers' preferences. In some robots, especially those used in spray painting applications, it is normal to find the actuator, the power supply, and the control system each being a separate module. In other applications the control system, power unit, and actuator are all a single machine. While the configuration can vary widely, each industrial robot must have all three basic components to function properly.

In order to operate an industrial robot, it is necessary to in some way teach the machine the motions you wish it to execute. This information must be stored in some form so that when it is time to execute the motions, the machine is capable of running the entire sequence of motions automatically. In order to understand the operation of industrial robots better, let us examine three functions—teaching, memory, and program execution—as they work on a number of different types of machines.

Pick and Place Robots

The easiest machine is the simple, non-servo, pick and place robot. Teaching the program to this type of machine requires two distinct steps. First, the stops on the various actuators must be set so that a proper end position of each actuator is achieved. These are normally mechanical stops held in place by a bolt or other securing device. These must be loosened with a wrench, moved to their proper position, and retightened. Once this is accomplished, it is then necessary to indicate to the control system the sequence in which the actuators are to be driven against the stops. This requires setting the control system. A typical control system for this type of machine might be a rotating program drum into which pegs are inserted. As the drum rotates, the pegs operate valves that cause the actuators to turn on or off. The cycle in this type of control system becomes one rotation of the programming drum. This system works very much like the old-fashioned music box, in which a small rotating cylinder with small spikes protruding rotates so that the spikes play a tune on a set of small metal rods. Obviously, in this system a "memory" consists of these small pegs remaining in the proper holes. Each time the drum rotates, it operates the same valves at the same time, providing a repeatable program. Normally the power source for this type of machine is air. Generally, this is the only type of industrial robot that can be operated efficiently utilizing air power, since with air it is difficult to locate the actuator without the use of external stops. Some of the larger and more sophisticated point to point non-servo robots utilize hydraulic power sources instead of air.

While the rotating drum type memory has been used extensively in controlling this type of machine, it is certainly not the only type of control system available. There are a variety of other control systems which can be programmed to turn on and off the various control valves in the proper sequence.

Point to Point Robots

Programming a point to point servo controlled machine is a bit different. These machines normally are programmed utilizing a hand held teach pendent. This teach unit allows the programmer to manually operate the various valves controlling the machine, although at reduced speed. Programming consists of moving each of the axes of the machine until the actuator is in the desired position to start the program. An "enter" button is pressed to record this position. Using the hand held teach pendent the programmer repositions the actuator into the second point of the program and again presses the "enter" button. This sequence is repeated until each of the points in the program has been recorded. When the machine is run, the actuator first moves to the first point recorded, then to the next, then to the next, and so forth until the entire sequence which has been programmed is repeated.

Once we turn to the servo controlled industrial robot, we are generally confronted with electronic control systems, normally computer controlled. The various points now are stored in the computer's memory as electronic signals.

Power sources for servo controlled point to point robots are normally hydraulic. There are several electric, direct current drive robots available. In an electric robot an electric motor operates the actuator through some type of gear train and linkage.

Continuous Path Robots

The third type of industrial robot, the servo controlled continuous path robot, is programmed and operated in yet another manner.

Since continuous path robots are designed to reproduce accurately complex, three dimensional paths in space, such as the motions of a spray painter painting an object, a new programming method is needed. What is required is a program comprised of many points very close together which trace out the path desired. Obviously, using the same teach pendant that is used with the point to point robots would be extremely time-consuming.

The method that has been developed is actually quite simple. The arm of the actuator is counterbalanced in some manner, normally by using springs, to make the arm weightless. The drive system is bypassed so that the arm can be grabbed and moved around by a person. The program on this type of machine is entered by depressing and holding a program button while moving the arm through the desired motions. The computer records the position of the arm many times a second. This enters thousands of points into the memory, each very close to the one just before and the one just after it. When operating, the robot moves through these points at approximately the same rate that the original program was input.

Again, in this system the memory is electronic, and because so many points must be stored, a much larger memory capacity is necessary on a continuous path robot. The power source for almost all continuous path type robots is hydraulic, since bypassing the power source during programming is easier in hydraulic systems than in any others.

Using continuous path robots for spray painting involves one additional consideration normally not a problem in other robot applications. Many, if not most, of the finishes which are sprayed today are both volatile and explosive. If these finishes are concentrated in the air in the right amount, the entire atmosphere becomes prone to explosion. Any spark or ignition source that should occur might set off an explosion resulting in damage and danger to personnel. For this reason, spray painting robots must be designed so that they do not, under any circumstances, provide this source of ignition. This is one of the major reasons why spray painting robots are generally powered by hydraulic fluid. The hydraulic servo valves can be operated on very low voltage and current levels, minimizing the difficulty in eliminating a source of ignition.

This source of ignition problem can be handled in two manners. The first, generally called explosion proof, is to house all electronic and spark producing electrical components within heavy, massive cast metal containers in such a manner that should a spark or other source of ignition ignite the atmosphere within the container, the resulting explosion can be contained and defused without flames ever propagating to the air outside the container. In practice, this is quite difficult to do for something as complex as a spray painting robot.

The second method consists of making the robot intrinsically safe. This means that regardless of what happens and regardless of what fails, a source of ignition cannot occur on the actuator. In order to do this, special electronic circuitry is required which acts as a very sensitive and very fast fuse. Any time a failure or problem exists which could cause a spark at the actuator, this circuit immediately suppresses the spark and removes the source of power. These safety circuits must also be redundant within the machine to provide the highest possible level of assurance that an ignition source cannot and will not be present.

Outputs

Using the appropriate teaching method, the programmer has been able to input a series of motions which the robot is capable of reproducing. This, however, is only part of what will be necessary for the robot to accomplish useful work.

Somewhere on the end of the actuator, the robot is equipped with a tool or hand. This hand may be a clamping device used for grasping a part which is to be moved. The tool may be a spray gun, hand drill, grinder, etc., as dictated by the particular application. In any case, a means must be present for turning the tool on and off. In the case of a gripper hand, it is necessary at some point in the program to instruct the hand to close and at some other point to instruct it to open again.

In order to provide a means for operating end of arm tooling, as it is called, an output must be present. An output is nothing more than an on/off switch controlled by the robot's control system. This switch turns the tool on and off in the same manner as a light switch in your own home turns a light on and off.

Most end of arm tooling can be operated by a single output; however, many industrial robots today are available with a variety of different output switches capable of operating not only end of arm tooling but also various fixtures, jigs, tools, etc. These outputs become a vital part of a robot installation and also provide starting signals for machines which are used in conjunction with the robot. For example, if an industrial robot has been set up to unload and reload a metalworking machine, a series of outputs will be required. Once the metalworking machine has completed its task, the robot must reach in and grasp the finished, machined part. An output will be required to close the gripper around the part prior to removing it. Another output will be required to unlock the automatic clamping device required to hold the part on the metalworking machine while it was being machined. Once the part has been removed and placed in the appropriate exit position, a second, unmachined part is gripped and placed within the automatic clamping fixture. The second output again clamps the part in place, and the robot hand gripper then releases it. Once the robot is clear of the work area, a third output is required to provide a start signal to the metalworking machine.

In the above example at least three outputs were required—one to operate the hand gripper, one to operate the clamping device on the metalworking machine, and one to start the metalworking machine. It is obvious from this example that a robot installation consists of more than just placing a robot in front of a machine tool and turning it on. The robot must be connected to and integrated with a complete system, including end of arm tooling, clamping devices, and the like.

It is obvious that at least one and in many cases several outputs will be required for a particular application. These outputs give the robot the ability to operate things within reach and provide the communication link whereby the robot is capable of talking to or instructing other devices within its working area.

Programming outputs is accomplished in essentially the same manner as programming motions. On a simple pick and place robot, the pegs on the drum programmer which turn on and off the various axes can also turn on and off various outputs. On the point to point robot the teach pendent utilized to move the actuator and program positions also can be utilized to turn outputs on and off.

Inputs

To return to the example of the metalworking machine, one other signal is necessary in order for this installation to function. When the metalworking machine has completed its cycle, and the part is finished, it must in some manner initiate the cycle of the industrial robot. On seeing the signal from the metalworking machine, the industrial robot must know that this signal initiates a program. This function is handled by a capability called "input."

An input signal is generally an instruction to the robot to wait until a signal is present before continuing to run the program. If the first step in a robot program is an input, the robot will wait until a signal is present, such as that from the metalworking machine, before proceeding with the rest of the program.

Input signals can also be used within a cycle to verify that a particular event has occurred prior to continuing the program. As an example, suppose we wish to verify that the clamping device holding the part in the metalworking machine is completely closed and locked prior to releasing the part and instructing the metalworking machine to begin cutting. This can be accomplished by providing a switch of some type on the clamping device which is only activated when the clamp is completely closed. Using the input to instruct the robot to wait until it receives a signal from the clamp before continuing insures that the metalworking machine tool will not begin cutting a part until it is clamped in place properly.

Another possible use for an input would be to verify that the finished part actually has been gripped and removed from the fixture prior to loading new stock. This can be accomplished by providing a switch against which the robot presses the part which has been removed from the metalworking machine. If the robot has failed to grip the part, there will be nothing in the gripper to press against the switch, thus preventing the robot from continuing.

It is obvious that in addition to multiple outputs, many robot installations will also require multiple inputs to function properly.

With some sophisticated industrial robots the inputs received can determine the actual cycle operated by the robot. As an example, let us return to the metalworking machine loading example. Suppose when the part was removed from the metalworking machine tool, it was placed in a special quality control fixture that automatically checked various measurements on the part.

An input signal from the test equipment can initiate a robot program instructing the robot to remove the part from the quality test fixture and place it in a bin of good parts. Should the quality test fixture, however, measure a parameter of the part that proves unacceptable, a different input signal to the robot could select a different program in which the robot removes the part from the quality control fixture and places it in a scrap bin. In this manner good parts can be separated from bad parts as part of an automatic cycle. This capability is called "branching" or "subroutining." In these systems the program can continue in a number of different directions with the direction determined by the inputs that are activated.

The inputs discussed here are the simplest forms available. More complex inputs can be developed which can react to various sensors located on either the robot or the machine tool, welder, etc., with which the robot is working. Very complex and capable programs can be developed using these techniques.

Obviously, the applications engineering necessary for these complex installations is more involved and the installations are more costly than other simpler applications. It is, however, necessary in many installations to provide this additional level of capability and flexibility in order to realize fully the economic benefits available.

Best Applications and Benefits

Again I must state that even with sophisticated input systems, branching, outputs, computer controls, and the like, those tasks which a robot can accomplish are quite limited when contrasted with a person's capabilities. When you stop and analyze it, a person's judgment and adaptive capabilities allow him or her to operate in almost totally unstructured environments. This is in contrast to the industrial robot, which must be provided with a consistent, predictable, and structured work place. This being the case, one may wonder whether the results and benefits of a robot installation are worth the efforts required. Another question which arises is the type of job for which robots are best suited.

In order to examine the best applications and maximum benefits of industrial robots, it is necessary to examine those situations in which industrial robots are superior to the people or systems they replace. As we have already stated, the robot does not possess superior adaptability or judgment. In addition, industrial robots do not possess the mechanical dexterity of a person's hand. They are not capable of the defined sense of touch, the compliance, the movement, or the flexibility of a person's hand. Robots do have the advantage, however, in their consistency. An industrial robot will perform its task with little or no variation, time after time, hour after hour, day after day, week after week. In contrast, a person presented with the simple, monotonous, repetitive jobs to which industrial robots are well suited, soon tires. Cycle to cycle variations creep into the person's work habits. The work pace becomes slower as the day wears on.

In examining the economic benefits of this robot consistency we find the real innate value of the industrial robot. The most obvious savings, of course, is in the cost of the direct labor which the robot replaces. This can be calculated easily by examining the direct labor rate for the job prior to the robot installation. However, it also must be noted that direct labor rates have been increasing at or just slightly behind the rate of inflation. In some major industries—the automotive industry, for example—direct labor rates have reached an excess of $15.00 per hour. During the recent rapid growth to this level, the operating costs including interest, programming, and maintenance for a typical industrial robot have remained nearly constant at approximately $5.00 per hour. While this expenditure for industrial robots may rise slightly over the next few years as the direct result of inflationary pressures, it is certain that direct labor rates will increase at a much, much faster rate as a result of continued inflation and population trends. This generally means that savings associated with a robot installation will increase year after year.

It must be noted, however, that the robot is not simply purchased, told where to report for work, and substituted for its human counterpart. There are certain expenses associated with the robot, including engineering; programming and program maintenance; and mechanical, electrical, and electronic maintenance. In addition, someone must tend or supervise the robot, even if only indirectly. In highly robotized, automated operations the general rule which seems to be developing is that one worker can service about five robots efficiently and adequately. These additional costs must be taken into consideration when evaluating the economic benefits of a robot installation.

There are, however, many secondary benefits of the robot installation, some of which may equal or surpass the direct labor savings. It is generally agreed that any installation in which a manual operation is added to an automatic cycle can benefit from robot installations. As an example, a sequence in which a machine tool is manually loaded, then automatically machines a part with manual unloading and reloading, will normally benefit by the use of robots. In this type of installation, the total time it takes to produce a finished part is the time it takes the machine tool to make the part plus the time it takes to unload and reload the stock manually. This unload and reload time can vary considerably, cycle to cycle and hour to hour. As the day wears on, the tedium of the job easily and understandably can lead to longer and longer cycle times.

The installation of an industrial robot, if properly done, will result in the shortest possible load and unload cycle. The real benefit, however, comes from the fact that the cycle time programmed is the same, time after time, hour after hour. This reduced overall cycle time can result in significant increases in machine production and machine capacity. The savings available from this type of improvement can be significant when the improvement increases the overall productive capacity of a very expensive machine tool. The expanded capacity or production capability resulting from the installation of industrial robots on a number of machine tools many times can result in increased capacity that otherwise might require the purchase of an additional expensive machine tool.

Here is just a note about some real world facts concerning people versus robots: It is virtually guaranteed that should you approach an employee to time study his or her capability of doing a job versus the capability of a robot, the employee easily will outperform the industrial robot. As I stated earlier, a properly motivated person can outperform a robot. That high level of performance, however, could never and likely should never be sustained by a person day in and day out. The types of jobs that industrial robots handle well are not the types of jobs that provide people with a high level of motivation.

An industrial robot's consistency of operation can also lead to more consistent, higher quality production from the equipment it services. As an example, in an injection molding machine, plastic material is heated in a barrel by a rotating screw, then forced under high pressure into a closed mold. In that mold the plastic material is cooled and hardens. When the mold opens, an operator, either a person or a robot, removes the finished part and initiates the next cycle. While the part is cooling, however, the plastic needed for the next cycle is heated. A person operating an injection molding machine will vary the time he or she takes to remove the finished part and close the safety gate to initiate a new cycle. This variation in time means that each cycle provides the plastic for the next part with a slightly different heating time. This means that each part may undergo a slightly different heating and cooling cycle, thus reducing the consistency of the finished product.

This can be overcome simply by allowing a longer than normal cooling cycle so that even when the part removal takes longer, a properly cooled part will be produced. The result is a longer than necessary overall cycle and less than possible consistency part to part.

It is obvious that the introduction of an industrial robot in this role will provide an exact part to part cycle which can possibly be reduced from that which occurs during manual operation. Increased quality may also result in decreased rejects. One of the best examples of this is in the slurry dipping process found in investment casting. Investment casting consists of making a wax pattern of a complex three dimensional part you wish to have reproduced in metal. This wax pattern is dipped into a slurry bath comprised of a number of liquid chemical binders. The wet wax pattern then is dipped into a fluidized bed of sand. A fluidized bed is nothing more than a large tub of sand which has air pumped into the bottom of it. The sand acts very much like a liquid and easily allows the wet wax pattern to be dipped and coated with sand. The sand coated pattern then is returned to the slurry dip and back to the fluidized bed of sand over and over, building a multi-layered sand mold around the wax pattern. This mold and wax pattern then is placed in an oven which cures and hardens the sand and melts the wax. The wax is poured out, leaving a female mold of the complex pattern into which molten metal can be poured.

As a manual process, slurry dipping results in a high level of reject parts since small bubbles tend to form on the surface of the wax mold when dipped into the slurry. Removing these manually is a difficult task and is never very efficient. However, a robot can spin the dipped, wax pattern at high speeds at the end of the arm, using centrifugal force to remove the bubbles even in the deep recesses of a complex part. The result of using industrial robots in this application is a dramatic and well documented increase in productivity and an almost total elimination of the rejects common in manual operations.

In some installations material savings may provide a highly significant cost savings. In many paint spraying applications the materials savings associated with the consistent cycle of an industrial robot can alone easily justify the cost of the installation. Spray finishing is an area in which working conditions are difficult, and fatigue can easily set in. Fatigue can result in less than optimal spray patterns and higher than necessary material usages. There is no question that an experienced spray finishing operator likely can match the performance of an industrial robot on a single part or even a few parts. It is, however, a documented fact that over the long run in daily production industrial robots can provide a 10%–30% material saving over even the best and most consistent spray painters. This reduced material usage also provides other less obvious advantages, such as reduced overspray cleanup, reduced spray booth maintenance, more consistent product quality, reduced rejects, and reduced rework.

I have been concentrating on the economic benefits of utilizing industrial robots. However, there are other reasons why a robot might be chosen to do a particular job. After all, robots are machines, and as machines they can be exposed to the dirty, unhealthy, or unsafe jobs which every day are being found unfit for personnel. Robots can reach into the jaws of fast acting, high pressure presses knowing that a malfunction will result only in mechanical damage and not injury or death. Robots can spray toxic or dangerous materials. Robots can operate in unhealthy or explosive atmospheres. In short, robots can provide the answer to how to perform many of the jobs that our government is now telling us are too dangerous, unsafe, or unhealthy to be performed by people.

In many of these installations, the choice is between providing the necessary safety devices, safety procedures, ventilation, special safety clothing, and gear or turning the job over to an industrial robot. It has been found in the past and will likely continue to be found in the future that robots provide the surest, lowest-cost, and best alternatives when faced with this choice.

From the few examples covered, it is obvious that many good, logical, valid reasons exist for the installation of industrial robots. It should also become apparent that each installation has unique characteristics, economics, and considerations that will determine the overall suitability of robots. However, when examining this justification be sure to take all facts into account—both the pluses and the minuses—as well as the effects of the installation on overall plant capacity, productivity, safety, and quality.

Chapter Three
JOBS FOR ROBOTS

One way to examine the capabilities and limitations of industrial robots is to examine areas in which industrial robots are being used or considered. In 1980 the United States had between 3,500 and 4,000 industrial robots in operation with an additional 2,500 or so in use throughout Europe. The Japanese claim to have in excess of 35,000 industrial robots operating; however, their definition of an industrial robot includes many devices not considered robots elsewhere. In actuality, the Japanese have at least 7,000 bonafide industrial robots in operation. The attitude which has led to this superior utilization of robot technology by the Japanese has contributed to their rapid growth in industrialization and productivity. By 1981 the pace quickened, with the United States producing between 1,500 and 2,000 robots in one year.

In examining uses for industrial robots within the U.S., I will classify robots into a number of categories. These categories reflect the basic function the robot performs, such as material handling, welding, and the like. I will then provide examples of each function being performed in various industries. A discussion of the basic requirements for a robot installation in each area will also be presented.

Material Handling

Material handling robots include all applications in which the robot is picking up, transferring, or moving parts. Probably the largest portion of material handling robots is utilized for machine loading and unloading tasks.

The automobile industry, which historically has been the leader in using industrial robots, uses a number of material handling robots. These generally are used for loading and unloading equipment, such as forming presses. One example of this type of installation is a company which manufactures frames for automobiles. Over a seven year period this firm has installed 13 industrial robots to load and unload forming presses. A payback period of six months justified this installation. In addition, the dangers inherent in having operators place their hands between the dies of the forming press were eliminated. This installation proved quite successful, although the engineering and start up efforts were much greater than the manufacturer had anticipated. The success of the installation, however, is borne out by the fact that this manufacturer is now seriously considering converting other factory operations to robots.

The tendency to underestimate the time, effort, and resources necessary to implement the first industrial robot is common. The peripheral work and engineering necessary to take the standard products offered by robot manufacturers and put them to work in a facility is not as simple as first envisioned. In most applications the job of integrating the robot with the equipment it must work with is quite extensive. In order to keep from being surprised by the efforts required, get a good, accurate estimate of the cost and time necessary for installation, and assume the first couple of installations will cost at least double and in some cases triple your first estimates.

In the die casting industry robots have found ready acceptance in removing parts from die casting machines. A die cast machine works very much like an injection molding machine for plastics. In die casting a closed, metal mold is filled with a molten metal which then is cooled and ejected from the tool. Just as in injection molding, the time required for the operator to reach in and remove the finished part is added to the total cycle time required for manufacturing. Robots in these applications have increased the productive capabilities of die casting machines by 200% to 300%. The additional capacity achieved in die casting is normally greater than that achieved in other processes such as injection molding. The reason for this is that cycle times in die casting are generally under 15 seconds, with many cycles in the 5 to 10 second range. The efforts of the operator to remove the part from the machine can easily become 50% to 75% of the total operation cycle.

Once parts are removed from a die casting machine, they are dipped in a cold water bath and then placed in a trim die which removes spurs, runners, and unwanted flash. This job can be extremely tedious when manned on a continuous basis. Industrial robots have found ready acceptance in this application.

Another common method of operating a die casting machine is to provide a water bath underneath the platens of the machine. As the mold opens, the newly formed hot parts are ejected and fall into this water bath. A feed conveyor then moves the parts from the bottom of the water bath up an incline and deposits them in some type of collector bin. Parts then are hand sorted and fed into the necessary trim dies and post finishing fixtures.

Even with this system, which is in itself quite efficient, justification for using robots can be established based on the elimination of the sorting and post finishing operations. Robots also provide for increased quality and decreased rejects of those parts which cannot be dumped together in a part holding bin.

Another material handling application in the foundry is the investment casting process described in the last chapter. It is estimated that foundry operations employ approximately 400 industrial robots today and over the last 10 years have boasted a growth in the number of installations by almost 50% per year. It is obvious from these figures that the major benefits outlined in the last chapter can have a significant impact in this industry.

Another material handling task which is performed extensively in industry is palletizing and depalletizing parts. Finished parts coming off a conveyor many times must be stacked or interwoven on a pallet. These parts then either are banded or shrink wrapped into a final package. By the same token, raw material for many industrial processes is delivered in palletized form. These parts must be unstacked and fed into the subsequent processes. Both palletizing and depalletizing, and in fact packing and unpacking parts in general, are tasks which can be handled by robots today. These tasks present the robot with additional complexities not found in other tasks, since each part on a pallet is located in a different position. Robots have been developed which are long enough to take into account the servicing of each individual position of a pallet. Techniques have been developed which instruct the arm to move in a particular direction until a contact or touch sensor on the hand comes in contact with the part. This system is especially useful in simple stacking and unstacking applications. Some of the more sophisticated robots have palletizing routines whereby the orientation of the palletized load, the number of layers, and number of positions on each layer can be indicated so that the robot can service each position on the pallet individually.

Palletizing and packaging operations provide a useful occupation for industrial robots, even though they may require a higher level of sophistication than other tasks in the material handling arena.

A new but potentially large application for material handling robots is in the production cell concept. In this system a number of machine tools are grouped around a material handling robot in such a way that parts can be transferred easily from one automated machine to another. These cells are designed so that a variety of different parts can be produced to a finished or almost finished state at one time. Material movement in and out of a production cell can be automated. Production cells can be the basic building blocks necessary for almost totally automated machined part production.

The machines which are grouped into a production cell can be varied according to the operations necessary to produce a particular component. The trend in manufacturing is to group parts into families based on the operations necessary to produce them. A production cell can be devised which can produce a number of different components which would be classified in the same family. With this system highly efficient and cost effective production can be achieved.

The typical configuration of a production cell consists of a variety of machine tools and quality control fixtures arranged in a circle with an industrial robot in the center. The industrial robot has the capability of reaching each of the machine tools and fixtures within the cell. Holding areas or short term storage areas are included so that production on most pieces of equipment can continue even if one of the machines should malfunction. Obviously from this description it is apparent that robots servicing production cells require more sophisticated programming, subroutining, branching, input, and output capability than those servicing a single machine. Production cell design, engineering, and implementation entail some of the most sophisticated robot applications in use today.

What are the basic requirements for a robot in a material handling application? The first, of course, is a robot which has a weight carrying capability greater than the heaviest part you wish to move. It is important to be certain that a reasonable safety factor is built in and that you are not operating the robot at its maximum capability most of the time. The reach and working envelope of the robot must be such that all the areas which the robot must reach are either accessible or can be moved so that they are accessible. A hand or gripper of some type must be developed that will handle the particular parts in question. While several companies now offer standard hands, grippers, and end effectors, it is necessary in most applications to design and develop these for the specific requirements.

The robot control system must have the capability of operating the hand or end effector. It must also have input/output capability so that it is capable of communicating with each of the machines it is to service. If the task includes packaging or palletizing/depalletizing operations, the control system either must have a large enough memory to build a complete pallet load program or must be able to service each of the positions on the pallet in some way. The level of safety interlocks must be determined. These interlocks make certain that various functions have occurred, such as verifying that the parts have been actually grasped, checking that the parts have been properly loaded or unloaded, and the like.

There will be substantial growth in material handling applications for robots in the next few years. These applications, especially those involving some of the machine loading and unloading areas, provide more consistent working environments within industry. This natural consistency of part orientation and placement makes these applications a natural for robots. At the same time these jobs are some of the more boring, routine, and undesirable from a worker's standpoint. You will find that these criteria—consistency in the work place and monotonous, tedious work—provide some of the necessary basics for industrial robots.

Welding

By 1980 welding was the major area of robot applications in the U.S. The use of industrial robots for welding applications is separated into two distinct categories. The first of these is spot welding.

Spot welding is used to join pieces of sheet metal together. Much of the assembly and construction of a typical automobile consists of spot welding the various sheet metal components together. As the name suggests, spot welding consists of joining two pieces of metal by laying them on top of each other and utilizing an electric spot welding gun to fuse the metal pieces together in a number of small, round areas or spots.

The second type of welding in which robots are applied is MIG welding. MIG stands for "Metal Inert Gas," which is a description of the process.

MIG welding is used to joint heavier sections of material than spot welding can handle. In MIG welding the two pieces of metal which are to be joined are butted together and electrically grounded. A thin metal wire is brought in close proximity to the butt joint and a heavy electrical current is passed through the wire. The spark resulting from the electric current passing from the wire to the pieces of grounded metal melts the end of the wire and fuses the edges of the metal parts together. As the end of the wire melts, additional wire is fed forward, continuously supplying additional molten metal to the seam.

The inert gas part of the definition refers to the fact that unless the area where the weld is being made is insulated by a gas other than air, the resulting weld will have a very low integrity. This inert gas either is pumped into the area of the weld or is generated by a coating on the wire which produces the inert gas as the wire melts.

Spot welding provides one of the largest markets for industrial robots today. Most automobile manufacturers utilize industrial robots for the spot welding of automobiles on a production line in at least one of their plants. The spot welding guns used to assemble an automotive body are fairly large and heavy. Even when counterbalanced by springs, they are burdensome for a person to maneuver. This fact, coupled with the short period of time available on an automotive production line to accomplish the necessary welding, can easily result in missed welds, inconsistent placement, and poor weld integrity.

The introduction of industrial robots to this application comes in two forms. In the first system the auto body is indexed to a welding area where a number of industrial robots move in and simultaneously spot weld various parts of the automobile together. The body then may move to another welding station where additional welds are placed, then to a third and a fourth and so forth. In this system the body maintains a stationary position as the robots move in and perform their jobs.

With the advent of line tracking, the necessity of stopping the automobile body to perform the welding has been eliminated. Line tracking is a sophisticated robot capability which allows a robot to work on parts as they move down a production line. The speed of the production line is measured by the robot's control system, and the robot arm is controlled automatically to adjust for variations in line speed. This system seems to be gaining wide acceptance by the automobile industry because it requires less specialized conveyor equipment than those installations which require the product to remain stationary.

There seem to be quite a number of successful European installations in production MIG welding. Production MIG welding has not gained widespread acceptance in the U.S., although the heavy construction industries in the U.S. seem to have a need for this type of equipment. Certain factors seem to be holding back this acceptance.

The major problem is that in most welding applications the various parts to be joined are not located precisely. The joint to be welded is not in exactly the same place time after time. The job the robot must perform is therefore not consistent cycle to cycle. For a successful MIG welding operation a consistency of mating parts, not normally associated with this production method, must be maintained. This requirement can increase the cost of production enough to reduce or eliminate any potential savings.

Another solution to this problem, which is being worked on diligently by a number of manufacturers, is to provide some type of sensor and adaption system on the robot. While welding, the robot will automatically compensate for variations in the parts being welded. A high enough level of sophistication and technology seems to exist today to make this goal feasible. It is only a matter of time before MIG welding becomes a very fast growing segment of the robot market.

MIG welding is an interesting application for industrial robots. While it is a tedious and tiring job, it is also a skilled job. This seems to be an instance in which robots are a substitute for skilled plant personnel. Even here, however, the driving force behind the use of robotics is the lack of an adequate and growing supply of skilled labor. There is little doubt that even in the future, those who develop a skill in this area will find a more than adequate market for their skills. As in other instances of robot utilization, it is the continuous, repetitive, boring production welding which is being automated, leaving the more varied and interesting applications for people.

There seem to be two distinct MIG welding technologies developing. The first of these involves machines capable of very, very accurate path tracking. On these machines smaller assemblies are fixtured so that the weld joints are positioned accurately and the welding head is moved accurately along the weld seam. Some of these machines even include programmable rotation and tilt on the welding fixture itself. Some machines, which resemble machine tools more than industrial robots, are capable of producing good, consistent, and inexpensive production welds. The control system operating the machine also controls the various functions of the welder. Systems of this type have been developed and are commercially available for production aluminum welding.

Another type of robot welder, typified by the Apprentice machine produced by Unimade, is designed for production welding in which weld seam positions are not consistent. In this system the torch is replaced by a programming wheel for initial program input. The wheel is rolled over the weld seam to input the program. Then the torch is replaced and the weld completed. In this system, when the weld is completed, the unit then is reprogrammed for the next weld. This system proves most beneficial for welding articles which are large. It also provides welds on products whose weld seams will vary greatly from weld to weld.

The first requirement in developing a robot spot welder is a robot with the capability of handling the large, bulky spot welding head.

The robot must be able to reach all of the necessary spot welding positions with the required accuracy. In addition, an output signal is necessary to operate the spot welding head. In spot welding it is not uncommon for one of the tips of the spot welding gun to stick occasionally to the sheet metal being welded. The robot in some way must be able to detect when this occurs and provide a shaking or rocking motion to break the tip loose.

In general, arc welding and MIG welding require a higher level of positioning accuracy than the general purpose material handling applications do. Not only must this accuracy be present at the end points, but the path that the robot takes between programmed points also must be highly predictable and accurate. Arc welding applications entail equipment which is more sophisticated and higher-cost than many of the other robot applications. This higher cost in both sophistication and investment pays off, however, in substantially increased levels of productivity and in dramatic improvements in weld consistency and quality.

Spray Painting

The best counts available indicate that by 1980 there were slightly over 700 spray painting installations worldwide, with approximately half of those located in the U.S. No single, dominant installation type or application seems to have developed, and it is likely that many of the installations to date are evaluation facilities in anticipation of future and more large-scale applications. Many of the spray painting robots available in the U.S. today are of foreign manufacture, although a number of firms, including the one with which I am associated, manufacture and market American-made spray painting machines.

Spray finishing applications require a broader and more all-encompassing technological capability than some of the other applications discussed. The success of a robot installation depends on the ability of the installation engineers to eliminate those variations within the workplace to which the robot cannot adapt. There seem to be more variables in a spray finishing application than in others. These variables occur in more engineering disciplines than most others.

As with other applications part placement must be consistent time after time. Since the path of the robot through the entire cycle now determines the quality of the job performed, variation in line speed and positioning, which might be tolerable in most robot applications, may not be acceptable in spray finishing applications. In addition, other factors which are normally not a concern enter in. The paint delivery system must be consistent cycle after cycle. Small variations in the spray pattern or the amount of material delivered—variations which are quickly and easily accounted for by the human operator—must be eliminated in automated operations. The lines, guns, mixers, pots, etc. used to spray the paint must eliminate every possible variation. The finishing material itself must be consistent batch to batch and day to day. Again, normal variations in material, viscosity, solids content (the amount of solid paint versus the amount of thinner), air pressure, paint pressure, and even paint temperature must be maintained at the same level day after day. Factors such as the air flow within a paint booth can cause enough of a change to make the installation unsatisfactory.

Obviously the complexity of spray painting applications varies from simple and straightforward to highly complex. Two distinct types of spray painting applications seem to be emerging. In the first, parts on either a flat belt or overhead conveyor are moved slowly past a spray painting robot which senses the part presence and sprays the part. In the second application type, parts are presented to the robot housed, masked, or otherwise fixtured. These fixtures, which are normally hand loaded, are locked into position before the robot, and the robot spray paints the parts while they are stationary.

As the complexity of spray painting systems increases, robots can be developed which spray a number of different colors through the same nozzle. The color can be changed automatically during program execution. This system works by continuously circulating paint from the paint can to the tip of the spray gun and again back to the paint can. To change colors, a metered amount of solvent is put into the paint line automatically, and the second color is inserted directly following the solvent. As the section of solvent circulates to the tip of the spray gun and back to the return line, the first color from the line is emptied back into the paint can. Just before the section of solvent reaches the paint can, the return line is diverted to a drain or other disposal container. Once the new color reaches the end of the return line, it is switched so that it circulates back into the new color tank.

This establishes circulation of the second color. As soon as the spray gun is activated, a color change will have taken place.

This procedure, as you can well appreciate, takes rather close timing and some reasonably sophisticated equipment. The equipment, however, is standard and can allow quite a number of different colors to be sprayed through the same spray nozzle.

Most industry finishing lines are designed to handle a variety of different parts loaded randomly on the line. In order for a robot to be able to cope with a randomly loaded line, it must have the ability to store a program for each part which might be loaded. It also must be equipped with some way of identifying the next part to be sprayed. This can be accomplished by having the person loading the line, type into a terminal the program number for the part being loaded. If the total number of parts is not large, simple electric eyes or limit switches can provide the necessary information. More complex installations may require highly sophisticated recognition systems, such as the bar codes which are becoming familiar in supermarkets, or even vision driven pattern recognition systems.

As you can see, spray painting can vary dramatically in complexity from very simple to highly complex, multi-color, random part applications.

Since spray painting installations seem to be more expensive, more complex, and more touchy than other installations, a natural question might be, "Why bother?" The answer is that finishing materials have increased in price dramatically as a result of the increased costs of petrochemicals. Automated robot finishing systems can save between 10% and 30% in material costs alone. Since the likelihood is that oil and oil base paints will continue to increase in price faster than most commodities, future savings from such installations can be substantial.

In addition, recent questions about the long term health consequences of working in an atmosphere containing many of the paints and chemicals now sprayed foretell a time when human exposure to such chemicals may be reduced dramatically or eliminated by law. Industrial robots provide the only other alternate means of production should this occur. Many of the paints, solvents, and chemicals sprayed today are considered pollutants, and the paint installations must secure EPA approvals prior to operating. As rules become tighter and tighter, some areas find it very difficult to meet the necessary emissions requirements while maintaining an atmosphere inside the plant fit for human occupation. Industrial robots in some of these areas can operate in an atmosphere so laden with poisonous or deadly fumes and chemicals that human beings could not survive. This can reduce substantially the amount of material exhausted to the outside atmosphere and bring otherwise non-complying installations into compliance.

As we can see, the benefits of utilizing spray painting robots are significant and unquestionable. Despite the fact that a dedicated effort is required for a successful installation, the growth of spray painting robots through the next several years should be substantial.

As more and more successful applications are developed, a certain level of expertise with spray painting robots results. This experience has resulted in an increasing level of confidence. In many applications the cost savings are obvious. As the cost of the equipment comes down to the point where a one year payback on single shift spray painting operation is possible, the use of industrial robots in painting will expand rapidly.

There are several elements necessary for a spray painting robot application. First a robot with continuous path capabilities will be needed, since all but the simplest and most rudimentary spray patterns require continuous path capability. The machine will have to be intrinsically safe. These two basic capabilities normally will be found only in robots designed primarily for spray painting. It is not advisable to take machines not designed as spray painting machines and attempt to spray paint with them. They will lack some basic capability such as continuous path tracking, lead through teach programming, intrinsically safe circuitry, output capabilities, or the like.

In addition to the robot, a spray gun and a system for turning the gun on and off must be devised. A multitude of suppliers and literally thousands of different systems can be used. This is where expertise in finishing equipment and materials is valuable.

If more than one part will be loaded on a line randomly, the robot control system must be capable of storing a separate program for each different part. It must also have the capability of randomly calling any of these programs. A system must be developed for indicating to the control system the program to be run. A signal must be generated indicating that the part is in position and that the program should start.

Most of the manufacturers or distributors of spray painting robots are capable of assisting in the development of this system. As time goes on, you will find more and more prepackaged systems comprised of a robot, a paint delivery system, the necessary mass memories storage, and part recognition systems.

Assembly

Approximately 85% of all manual labor expended in industry in the United States could be classified as assembly. While we could consider welding an assembly process, for the purposes of this section we will consider assembly to be those processes in which products are brought together, fitted together, screwed in place, or otherwise assembled.

It is only in the last few years with a joint effort between the auto industry and Unimation, Inc., a subsidiary of Condec and the first manufacturer of industrial robots, that the first assembly robots have been developed. These very fast, jointed arm designed robots with very light (less than 5 pounds) load carrying capabilities have set the stage for what may be a whole new era in robot utilization.

Several major electronic and computer companies have, on their own, developed and implemented the use of similar assembly robots on their production lines. Conjecture today is that these large, well financed and well managed high technology companies soon will enter the robot manufacturing business, supplying the same type of assembly robots to industry.

These machines are capable of doing light-duty, very precise assembly of various components on either a cycling or moving assembly line. Accordingly, many assume that assembly lines soon will appear which have the simpler assembly tasks handled by robots and the more complex tasks handled by persons working alongside robots.

In addition to industry's efforts in this direction, several research organizations associated with major universities have continuing programs under way in developing robot technology and techniques for use in industrial assembly. Two notable examples of this effort are Stanford Research Institute, which has had a continuing project concerned with the use of sight systems in assembly, and the Charles Stark Draper Laboratories, which has developed a device called the remote center compliance hand which allows very smooth, rapid insertion of parts which mate together very cleanly.

One very large company that has publicly indicated a major commitment to robotics conducted a study of assembly labor in its various plants. In that study it determined that some $900 million per year was being expended for tasks which could be defined as assembly. Of this, $600 million per year was being expended for jobs which could potentially be handled by robots. Of that $600 million, $300 million was being expended for tasks which could be handled by robots which were standard, available, off-the-shelf items in 1980.

In just one company the enormous potential of assembly robots can be seen. Just imagine the overall potential for this type of machine throughout industry where, as we stated earlier, 85% of labor is expended for assembly tasks.

It is generally understood and accepted that the extensive use of robots in assembly tasks is probably several years away. It seems to be an equally accepted fact that this application may very well be one of the most dramatic, if not the most dramatic, use of robots in industry.

It is against the use of assembly robots that the anti-robot, anti-automation forces have taken their stand. The claim is made that robots in this application are no longer relieving a person of hot, tiring, or difficult tasks. The assembly robot, in fact, operates in the same air-conditioned, well-lit assembly room where people sit comfortably doing their jobs.

This argument is further justified because the sheer size of the overall market for assembly robots could ensure a reduction in robot manufacturing costs far below present levels. An industrial robot could cost half or less the annual salary of the typical worker. The danger, they argue, is very real to any person accomplishing a task which now or at some time in the future could be taken over by an industrial robot.

These arguments might make sense except for the fact that they are based on a premise which, if not totally false, is at least very shortsighted. That premise is that, except for the change in the use of industrial robots, future society and future industry will remain unchanged. This premise has never been true at any time since the advent of the industrial revolution and is not true today.

Should the wholesale use of an assembly robot occur, and there are as many factors against such a possibility as for it, the greatest likelihood is that society would quickly adapt to the change, benefit from the advancement, and continue onward.

The same fear of massive unemployment was expressed in the early '60s as computer prices began to fall and their use became widespread. The same people who argue against the use of assembly robots—or any robots for that matter—argued then against the use of computers because of the massive white collar unemployment that would result as the faster, more accurate computer systems took over day to day clerical tasks. However, they overlooked the sociological change that additional technology brings with it.

What really happened was that the abilities of computers to provide information increased the desire and need for information. Rather than reducing the number of white collar workers, computers increased them. The overall amount of processed information increased at many, many times the employment rate, but the facts are the same. Higher levels of automation and technology provided better business systems, a higher standard of living, and increased, not decreased, employment.

Throughout history, increased productivity and automation have always resulted in increased employment and a higher standard of living. In contrast, those societies which have allowed the negative thinker to dominate and who have tried to save jobs and reduce unemployment by slowing and stifling innovation and progress have suffered the consequences of high unemployment.

There is no basis, therefore, for believing that the use of assembly robots will result in anything other than increased productivity, an increased standard of living, and increased employment.

Small part, light assembly robots seem to be the first type being developed. These are characterized by relatively high speeds, light payloads (5 pounds or less) and high accuracies. Assembly applications can quickly become complex and engineering oriented. Unless you are well versed in the application of robots and the theory and science of assembly engineering, using robots for assembly would not provide a good direction for initial robot installations. Most of the successful assembly applications have been installed by well-qualified, well-staffed, and experienced organizations, in some cases working with extensive government funding. Although the use of robots in assembly holds a great deal of promise, it does require a high level of understanding, engineering expertise, and experience.

Grinding, Sanding, and Machining

When we speak of machining or grinding, we generally think of machine tools which very accurately machine, cut, grind, or polish materials. In industry, however, there are a large number of machining jobs which are done by hand. Jobs such as sanding, buffing, polishing, deflashing, relief and radius grinding, and the like are performed each day by hand. Although industrial robots generally are not precise enough to accomplish tasks in place of machine tools, in many instances they can directly replace hand labor. As an example, if a person is hand routing a piece of wood using a guide fixture, a robot will not be able to grab the part and precisely route it without the fixture. It can, however, use the same fixture that was being used by hand and perform the routing operation.

Deflashing in which the grinding or deflashing tool is guided by the body of the part can also be accomplished. Recently the United States Air Force funded a project in conjunction with a major aerospace company to demonstrate the feasibility of industrial robots to utilize a typical aerospace hand drill fixture and automatically drill the sheet metal skins of aircraft. This application has proven successful and may provide the basis for increased use of industrial robots in this application. Despite the successes there has not been a major move within the U.S. to utilize industrial robots for machining tasks. Their use in Europe, however, is more prevalent.

While I have attempted to present in this chapter a bit of the flavor of industrial robot installations, it should be noted and understood that industrial robot applications are as varied and individual as the processes and products of the companies which employ them.

Chapter Four
WHERE DO WE START?

Now that you have an idea of what industrial robots are, how they work, what they do, and what tasks others are using them for, it's now time to turn to your own facility. The question arises, "Where will this equipment benefit me the most?"

As I stated earlier, robots are nothing more than a new form of industrial automation. To be fair to yourself in seeking uses for this automation, you would be well advised to examine your facilities on a general basis. While examining the facility to determine the benefits available from today's technology, you should take into account all forms of automation, not just robots. For many firms the excitement and romance of the word "robot" introduced them to the world of automation; however, they might initially benefit far more from other types or forms of automation.

I would recommend an examination, on a very objective basis, of your entire production operation. If your facility already employs various types of numerically controlled or computer controlled machines, such an examination may not be quite as beneficial as it will be in those firms where programmable automation is yet to be used. If you fall into the latter category, a thorough examination of your facility to determine the potential benefits of using modern-day automation may prove very profitable indeed.

There are several ways to go about determining the potential applications for automation within your firm. The first is to hire one of the experienced robot consultants. These people provide, for a nominal fee, an overview of the potential for robotic automation in a plant. This analysis normally requires a full-day tour of your plant with a knowledgeable guide, followed by one or two days of analysis and report preparation. The result of this effort can be quite enlightening, especially to those firms without internal automation engineers.

These same consultants are available to handle the design, implementation, and integration of robots into your facility. However, these services entail higher engineering fees. The use of knowledgeable consultants is recommended for those firms lacking the special technical expertise required by automation.

One word of caution. As the popularity of robots increases, it is likely that opportunists will be popping up everywhere disguised as robot consultants. Be certain that the organization you work with is qualified, capable, and above all experienced in robot application before you allow it to affect your operations.

A second method of determining where robots might be applied is by contacting the robot manufacturers. A personal call to the chief marketing executive of a robot manufacturing firm, in which you express a genuine interest in robots and indicate that a successful installation could result in reasonable future business, might very well bring a visit by an applications engineer. Even a visit by one of a manufacturer's sales personnel can prove quite helpful in pointing you in the proper direction. There are, however, several precautions to keep in mind when working with robot vendors. First, it should be understood that the robot market is growing very rapidly and has generated a great deal of industrial interest. Most robot manufacturers are completely buried in inquiries from interested potential customers. In order to get any special attention from these manufacturers, you must set yourself apart from this multitude of individual requests. The best way to do that is to make a commitment to robots and act accordingly. Provide the manufacturer with assurance that 1) you are serious; 2) sales to you can be of reasonable size; 3) you are ready to move quickly; and 4) you will work with those who assist you best.

The other precaution to take when using a robot manufacturer for engineering advice is that these people probably will provide you only with information on those applications for which they have suitable equipment. They may even recommend their equipment for a task which a substantially lower-cost, competitor's machine would perform quite adequately. Again, you can obtain a good deal of useful insight and information from robot vendors, but remember the motivation of those to whom you are talking.

The third way of evaluating your potential need for robots or automation is to have you, or someone in your organization, analyze your own operation. This sounds like the simplest of the three methods described. It is, however, actually the most difficult and the least likely to succeed. The reason for this is understanding. You and your people have a thorough understanding of what is happening within your organization. Even those processes and operations that would look absolutely ridiculous to an outsider, you justify with clear, precise logic. This understanding becomes a major impediment to gaining the objective view of your facility so essential if any useful insights are to result.

Knowing this, if you would still like to try to assess your potential need for robots, I will try to provide you with some guidelines. These guidelines originated as a simple guide for salespeople in our company. They were designed to find applications for a type of automation other than industrial robots; however, they apply equally well to robot installations.

The guidelines provide things to look for during a tour of your production operations. Before you begin your tour, I must again emphasize that you should attempt to observe what is going on within the operation objectively and analytically. Your observation is not intended either to endorse or condemn the operations, but is designed simply to identify opportunity. Be certain not to attempt to make a determination during this initial analysis of whether or not robots can perform various tasks. Any knowledgeable person easily can find many reasons why a particular operation cannot be automated. This thinking must, at this time, be avoided at all costs.

Crowds

The first tell-tale sign that improvements are possible is a crowd. What a crowd is will vary from plant to plant, but anytime there is an unusual concentration of people in one area, the potential for using automation may exist.

It is important to note, however, that improvements are not necessarily needed in the area where the crowd is. A process before or after the location of the crowd may be causing the excess labor usage and may need to be corrected.

Machine People

Machine people are production people who imitate a machine. Whenever you see capable workers performing tasks that are unchanging and that require no judgment or thinking, you have found a machine person. These machine people can be loading and unloading machines. They can be assembling a product repeatedly. They can be spray painting parts coming down a line. In short, they can be performing any one of thousands of jobs which are tedious, boring, and repetitive. Since machine people are operating like machines, they obviously can be replaced by machines and should be.

For reasons discussed in more detail in the section on labor relations, these people should be reassigned or retrained for jobs which are more demanding, more challenging, and more interesting than those from which they were removed. The new jobs can make fuller and better use of these people's talents and judgment, leaving the repetitive, unchanging jobs to robots.

Few industrial plants exist today without a liberal scattering of machine people. Machine people should be one of the most sought after areas during your tour, since they provide an opportunity for some of the quickest and easiest robot applications available.

In-Process Mountains

If your facility has large amounts of in-process and semi-finished goods, it is likely that both the opportunity and justification for further automation exists. In-process materials are quite expensive. They require clerical and management time to plan and direct their movement. They require large amounts of valuable floor space. They require large investments in material and labor costs. They should, in fact, be eliminated wherever possible.

In pursuit of this goal, two attacks work very well. The first of these is substituting multi-purpose, multi-function machine tools for individual operations. In a multi-function machine tool, parts which normally would require processing on two or more individual machines are processed by a single machine.

If you examine the labor content of a part which is processed through several machines with in-process storage between machines, you will find that in excess of 80% of the labor expended is worthless. "Worthless labor" is any effort for which you pay but which does not increase the value of the part. For example, removing a part from a storage bin is worthless labor; clamping the part in a drill press is worthless labor. However, drilling a hole in a part is useful labor, because it increases the value of the part. Removing the part from the drill press is worthless, and so is placing the part back in a storage bin. It is easy to see that in this simple process of drilling a hole in a part, the largest portion of the labor could be eliminated without affecting in any way the form or function of the part in question. It would only be necessary to drill the hole needed at some other point in the production cycle, when the part is clamped in another machine for another purpose. With that simple change, the part's value is increased without expending additional worthless labor.

Many new machines exist today which are capable of performing multi-machining functions on a part during a single load cycle. This eliminates both the worthless labor of loading and unloading a number of machines, as well as eliminates the need for in-process inventory between the various single purpose machines.

If several machines are necessary to finish a component properly, the work cell arrangement might prove attractive. Each of the necessary machine tools can be arranged around a centrally located industrial robot. This industrial robot then can move components from machine to machine, accomplishing a number of tasks at one setting and eliminating the necessity for large in-process inventories between the various machines. This concept can provide substantial savings in any operation where in-process mountains exist.

Closets With Skeletons

Imagine an important and prestigious visitor has requested a tour of your plant. During the tour there is one part of the plant which you avoid like the plague. That's your "skeleton closet."

Skeleton closets come in many shapes and many forms. They may have hot, dingy, dirty working conditions. They may involve primitive processes. They may even have unhealthy or dangerous working conditions. Regardless, almost every facility has at least one skeleton closet, and this area can provide fertile ground for improvement through automation.

Robots can work in dirty, unhealthy, and dangerous occupations. Their use can change situations you are ashamed of into robot applications you can be proud of. Since skeleton closets come in so many different forms, the robot applications necessary to improve them are also quite varied.

Ridiculous Processes

Here is where the real test of your objectivity comes. Few, if any, of the processes within your organization will look ridiculous because you understand their necessity and significance. These same processes, however, would look, if not ridiculous, then uninspired through the eyes of an objective, outside viewer.

Some examples might help. Have you ever walked through a plant and seen people in a room spray painting, routing fiberglass, or doing other tasks in space suits? The suits are not exactly space suits; they are completely enclosed suits with outside air pumped in to insulate the wearer from unpleasant working conditions. If you understand the poisonous nature of some of the spray materials or the long-term health risks in some of these jobs, the space suit might not seem ridiculous. However, an outsider comes away with the impression that there certainly must be a better way.

Do you have people within your plant jumping up and down on parts or machines, hitting, slapping, or kicking equipment to make production happen? Ridiculous processes.

I have seen finishing plants where buckets of parts were dunked into a paint tub, then removed by hand onto drying racks. Paint usage was unbelievably high, quality terribly low, and labor content higher than necessary. Another ridiculous process.

Try to look at your facility objectively. Although ridiculous processes are the hardest for insiders to see, they also provide some of the best potential for increases in productivity and some of the highest cost savings when properly automated.

Simple Solutions That Don't Work

There are times in production when the simple, straightforward methods have serious drawbacks. Examples of this are router bits in hand operated trimming that lasts 20 minutes or less; simple guide fixtures that take longer to locate and position than to use; simple, automatic spray guns which move back and forth but use 40% more paint than manual applications; hot parts which must be moved from a press to a quench tank, using gloves and safety equipment which cost so much and reduce output to a snail's pace. The list of simple solutions with serious drawbacks is almost endless.

Many times the simple and preferred process can be replaced by a more complex, automated, but much more profitable installation. This is a tricky area when it comes to cost justification. However, the potential for increasing productivity and decreasing cost is there.

Expensive Solutions

Whenever a firm is contemplating a large capital expenditure, especially for single purpose equipment designed to solve a simple problem, the potential exists for increased flexibility and decreased cost using industrial robots. Many times a flexible industrial robot can be substituted for fixed, single purpose automation with no loss of production volume. Robots, especially the general purpose type, can be reprogrammed to perform some other function should the task change. This is not true of fixed, single purpose automation, which is useless once the single task for which it was designed and built is eliminated. Any time a fixed piece of equipment is contemplated to perform a limited job function, the use of an industrial robot for that same job should at least be considered.

Bulges in the Line

Potential for automation also exists whenever a production line seems well balanced except in one area, where a task seems to require an inordinate amount of effort and personnel. Plastic molding machines or forming machines which need only a few people to efficiently produce thousands of parts, then send these parts into a large room where multitudes hand trim or hand deburr them, is an example of such a situation. In fact, whenever the inherent efficiency of a process is counteracted or nullified by the inefficiency of a post-process task, a very real opportunity exists for further automation.

This type of problem within industry is more commonplace than you might imagine. The manufacturers who supply some of the equipment that produces parts at high efficiencies and high speeds have one tiny problem that they haven't quite solved. This tiny problem can be handled easily by post-process work, so they claim. Hence, the bulge is born. Eliminating bulges in a production line of this type can be a very rewarding and profitable career.

Jobs You Wouldn't Want To Do

Whenever you see any employee performing a job that is so difficult, tedious, tiring, or boring that you would never consider doing that job yourself, you have found a likely candidate for further automation. Your employees are not much different from you. Those things you find undesirable, they also find undesirable. Those things you find boring, they also find boring. By eliminating these types of jobs, which are, by the way, the easiest for robots to perform, you can better utilize those talents your employees possess and realize much greater value for your labor dollar.

Salvage Yards

If you notice large amounts of scrap or substantial areas within a plant devoted to rework, an opportunity for automation exists. Automation can be justified easily in these situations, since the job can be done correctly and automatically the first time.

And so your tour ends. With a little effort, you have found some promising areas within your facility for the use of industrial robots. The tour, however, is only the start. It is likely that many other opportunities, more subtle in nature yet nonetheless quite profitable, can be found.

A normal result of the tour you have just taken is a rather deep level of despair. There is so much that can and needs to be done and so many opportunities that the implementation job seems almost impossible. This is a good time to understand that the task before you not only looks impossible but probably is, that is, if you attempt to do it all at once.

Learning to use robots is exactly that—a learning process. As in any learning process, if you wish to be successful, you need to take things slowly and carefully at first. As you gain confidence, capability, and knowledge, the task of implementing industrial robots and using them to the maximum benefit will become easier and easier.

As you have seen, many opportunities exist within your facility to utilize industrial robots, and a huge potential exists for you to become more competitive, more productive, and make higher profits using them. On the other hand, running headlong into the use of industrial robots can easily prove disastrous, with the real losses being not the monies expended but the loss of productivity and profits that successful installations would have brought.

Industrial robots have several characteristics not possessed by other machines. One is that they invoke emotional responses not just from the hourly paid employees but also from middle and upper management. Ignoring labor relation considerations can easily lead to robot installation failures.

An overly enthusiastic acceptance of the concept of using robots can also prove to be dangerous. Hoping that industrial robots will solve those problems which manual labor and conventional management techniques have been unable to solve is asking for trouble. The one thing that successful industrial robot applications require above all else is a large quantity of common sense.

Chapter Five
WILL IT WORK?

You have toured your facility and identified various opportunities available. The question naturally arises whether any of these areas can be automated using robots. Since the variety of tasks in manufacturing plants is so large and the same tasks within different plants vary, we will try to provide the basic guidelines which can determine whether or not a particular task is suitable for robot operation. The questions and criteria should be examined carefully in light of your knowledge of your present production. They should provide you with a method whereby those areas which are ripe for robot automation can be separated from those areas which today's technology cannot readily improve.

Can You Benefit from Increased Consistency?

Consistency is the one great attribute of a robot. As stated earlier, robots do not have the adaptability, overall dexterity, or the judgment of a human being. However, robots do surpass human beings in their ability to perform a task repeatedly in a precise, consistent manner. Any task that requires consistent, precise operation is one that is particularly attractive for robot automation.

The effects of highly consistent operation, at times, can be quite subtle. In order for an accurate assessment to be made, the question must be carefully approached. In certain processes improved product quality and reduced rejects are the natural result of a highly consistent and dependable production operation. When this is the case, justification for the robot installation is quite easy, and cost savings can be substantial.

Consistency of operations can also mean substantial material savings. This is especially true in spray painting applications, where savings of 10% to 30% are common. Once a program is established and perfected, it can be repeated consistently hour after hour, day after day. The corresponding consistency of a human being deteriorates as fatigue sets in. This produces products which may vary through the day utilizing more than the necessary amount of finishing material.

At times the materials savings take on a very subtle nature. One application of this type consists of a finishing line for a business machine housing. Prior to putting a robot in this application, the housing was painted using a reciprocating gun and a moving conveyor belt. As the part passed down the conveyor belt, the gun cycled back and forth across the belt. As the part came in place, a switch was tripped that turned on the gun. The gun cycled back and forth, placing an even coat of material across the entire part. This does not look like a process that can be improved using an industrial robot. However, the part contained a hole in the center which comprised 25% of its entire surface area. In addition, only a small section across the bottom of the part required a 3 to 4 mill paint coverage, since it was a wear surface. The remainder of the part would have an adequate finish with 1 to 1-1/2 mill coverage. The reciprocating gun painted the entire part with the same 3 to 4 mill coverage and also painted the 25% blank space in each part. By installing a robot that painted only the part and not the hole and placed the material with the proper coverage (1-1/2 mill over the entire part and 3 to 4 mill across the wear surface), a substantial material savings was achieved.

The ability of an industrial robot to provide a fast, consistent cycle can increase the overall productive capability of the machine tool or press which it is servicing. This is especially true when the load and unload cycle is added to the actual machining cycle. In this instance, the faster the parts can be extracted and loaded back into the machine tool, the greater the overall production level. Robots are capable of sustaining speeds and levels of production which would quickly fatigue a human being. Substantial savings in production costs and deferred new capital equipment expenditures can result from the utilization of industrial robots.

How Complex?

I have warned of the dangers of trying to automate processes which require robots that are excessively complex. Installations requiring complex sight, tactile senses, large computers, or highly complex operating programs should be avoided. Although the technology to perform these feats exists today and has been demonstrated in laboratories and trade shows, it is doubtful that such high technology products are practical in the average production shop. The cost of this special capability is rather high, and the training and expertise necessary to operate and maintain the equipment normally is beyond the capability of the average user. Another factor is that few shops exist today that do not have a multitude of simpler and more economical robot applications. These simpler applications not only cost less to install, but can provide more than adequate paybacks to insure a very respectable return on investment. Concentrate on the simpler applications and avoid extremely complex installations.

I do not wish to suggest that the capabilities I have advised avoiding do not or will not ever have a place in industry. I believe, in fact, exactly the opposite. Within the next few years capabilities which are now being demonstrated in the laboratories will be finding useful and justifiable applications within industry. I do believe, however, that any one reading this book is not now in a position to consider seriously this high level of technology. There is a natural tendency for those not familiar with robots to request the robot and robot manufacturer to supply all of the answers necessary for a successful installation. It is quite easy to ask for a sight system so that it becomes unnecessary to maintain part orientation. Attempts to take the easy way out and substitute super high technology for adequate planning have proven unsuccessful in the past and will prove the same in the future.

There is another side to the complexity issue. Applications which are very simple deserve approaches which are very simple. Certain applications are so simple that a simple guide, part orienter, or inexpensive fixed mechanism may be a better and more economical answer to the problem than the installation of an industrial robot.

In summary, when viewing the complexity of the proposed function to be automated, avoid using robots in those applications that are either very simple or very complex.

What Is the Production Rate?

In our definition of an industrial robot we indicated that industrial robots perform tasks normally performed by people in essentially the same manner as a person would perform them. To this can be added the fact that robots perform the tasks at approximately the same speed as a person. While there are exceptions to the above definition, it can be assumed that a robot will not operate at an appreciably higher speed than a person. Industrial robots can sustain production levels longer than a person can, especially when handling heavy, hot, or difficult materials in hot, dusty, dirty, or otherwise undesirable environments. Even with this, installing one industrial robot to perform the tasks normally performed by two or three people at the same speed that the two or three people performed them should be avoided. A good rule of thumb is that if a person is not capable of performing the task, even for a short period of time, at the production rate expected from the industrial robot, you are inviting trouble with the installation.

Small non-servo pick and place robots handling parts weighing only a few ounces can move at high speeds. Time is required, however, to grasp an object, move it, and release it, so that even when these relatively unsophisticated robots perform very simple tasks, cycle times of less than 3 to 5 seconds are difficult to sustain. Once either the complexity of the task or the weight of the items being carried increases, cycle time also increases.

Some servo type robots can operate at high speeds (up to 60" per second); however, when operating at these high speeds, repeat accuracy is diminished. When a part is to be either located and grasped or placed in a position with any degree of accuracy, the cycle must include time for the robot arm to decelerate and time for the part to settle, perhaps as much as 1 to 2 seconds with heavy loads. All these factors tend to slow the overall operating speed of the robot. As a general rule, always assume that robots operate at about the same pace as a person.

How Ordered Is the Environment?

Robots require a high degree of order and consistency in the working environment in order to perform successfully. Trying to operate a robot when parts are not consistently placed or major variations in the work place exist will be difficult or impossible.

On the other hand, few work places exist today which are sufficiently rigidized to allow a robot to simply be placed in position and started. Locating fixtures, holding conveyors, orientation and feed devices, and the like will need to be added to the industrial robot in order for it to perform useful work. A good deal of development is underway to make robots more flexible and adaptable to work in a bit more disoriented world than today. These developments will make the use of industrial robots easier in the years ahead. For the foreseeable future, it is generally more economical and practical to organize the work place for robots properly than to try to build into the robot the ability to handle a more disorganized work place. Many of the techniques necessary to properly hold, store, orient, and feed work pieces are well established. An individual or group with experience in this area can guide you in the proper direction.

I don't mean to suggest that there is no place at all today for sensor equipped robots. Vision or touch sensors, for example, may make sense in certain applications, such as locating a part on a conveyor belt or locating the top of a stack of pieces to be unloaded. The capability of this technology is still quite limited, and using ultra-sophisticated robotic systems to make up for a disoriented work place should be approached with extreme caution.

Is the Production Run Size Correct?

If only a small quantity of a product is going to be manufactured, the expense and effort required to utilize robots in their manufacture might well be greater than the expense necessary simply to produce the parts using manual labor. I do not recommend using robotics unless a part is going to be produced over and over and the programs, fixtures, hands, etc. necessary can be properly utilized. The only time that this rule doesn't apply is in health related applications, where it could prove dangerous for a person to perform the task even one or two times. In these instances, it very well may be feasible to utilize robots even to produce a single item.

Hydraulic, spherical coordinate, non-servo robots can function in a variety of industrial applications. *(Photos courtesy PRAB Robots Inc.)*

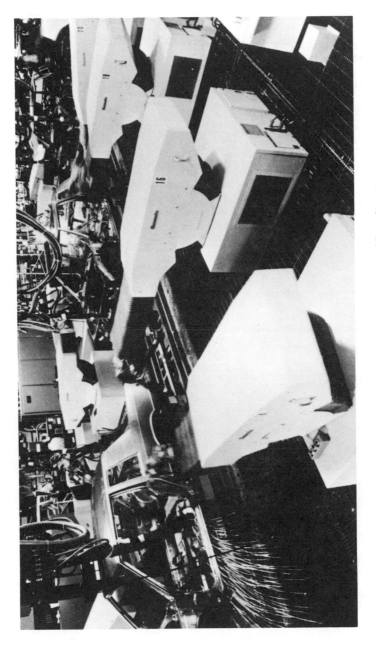

Hydraulic spherical coordinate, servo robots working on a spot welding line in the automotive industry *(Photo courtesy Unimation, Inc.)*

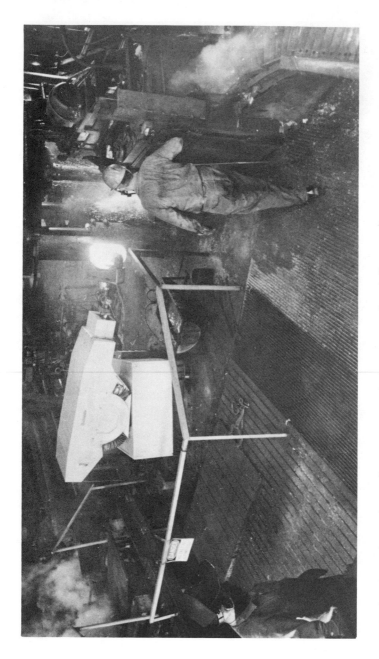

A hydraulic, spherical coordinate, servo robot moving a hot billit in a forging operation *(Photo courtesy Unimation, Inc.)*

A hydraulic, spherical coordinate, servo robot unloading and quenching a die cast part (*Photo courtesy Unimation, Inc.*)

Two small, air powered, cylindrical coordinate, non-servo robots for small part transfer *(Photos courtesy Seiko Instruments U.S.A., Inc.)*

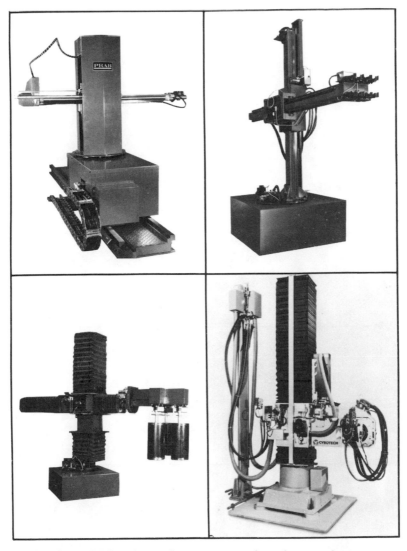

Hydraulic, cylindrical coordinate, servo robots for transferring parts weighing from 100 pounds (*upper left*) to 600 pounds (*lower left*) (*Photo courtesy PRAB Robots, Inc.*) *Lower Right:* A large, hydraulic, cylindrical coordinate, robot capable of handling loads of 175 lbs. (*Photo courtesy Cybotech Corporation*)

A large, hydraulic, cylindrical coordinate, servo robot capable of lifting 2,000 lbs. *(Photo courtesy PRAB Robots, Inc.)*

An electric Cartesian coordinate, servo robot for arc welding; the part positioners shown on each side of the robot also can be servo controlled. *(Photo courtesy Advanced Robotics)*

A small, electric, Cartesian coordinate, servo robot *(Photo courtesy Anorad Corporation)*

A large, hydraulic, jointed arm, servo robot for welding and material handling applications *(Photo courtesy Cincinnati Milacron)*

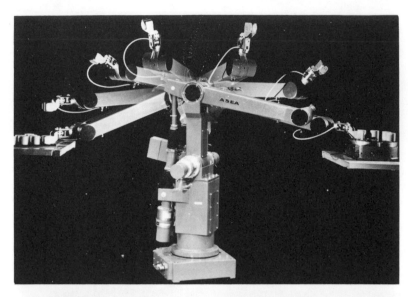

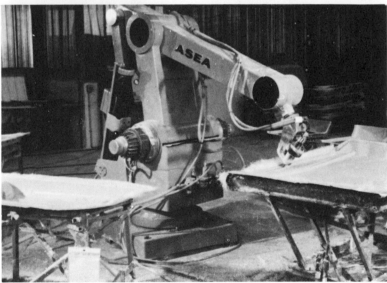

Top: Multiple exposure showing movement of an electric, jointed arm, servo robot; *Bottom:* Electric, jointed arm, servo robot trimming fiberglas parts *(Photos courtesy ASEA, Inc.)*

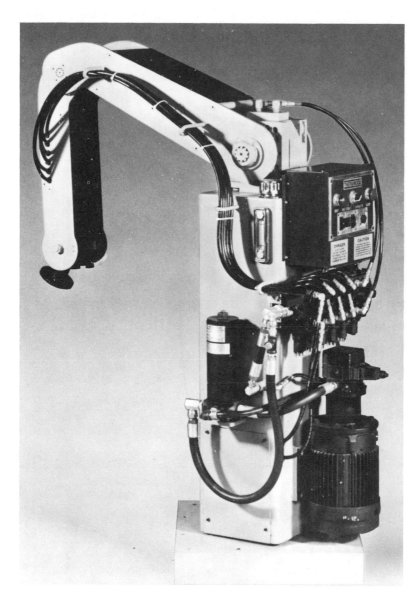

A hydraulic, jointed arm, servo robot for material handling *(Photo courtesy Thermwood Corporation)*

Electric, jointed arm, servo robots inspecting auto body dimensions *(Photo courtesy ASEA, Inc.)*

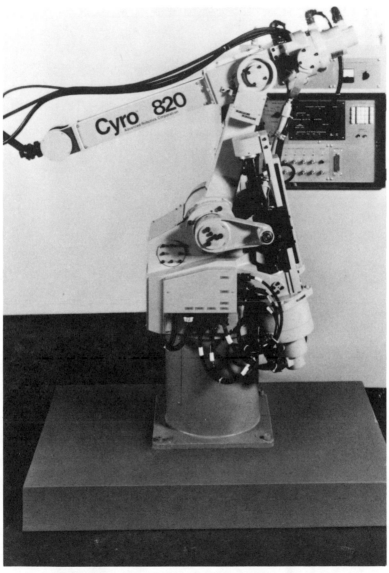

An electric, jointed arm, servo robot for use in arc welding *(Photo courtesy Advanced Robotics)*

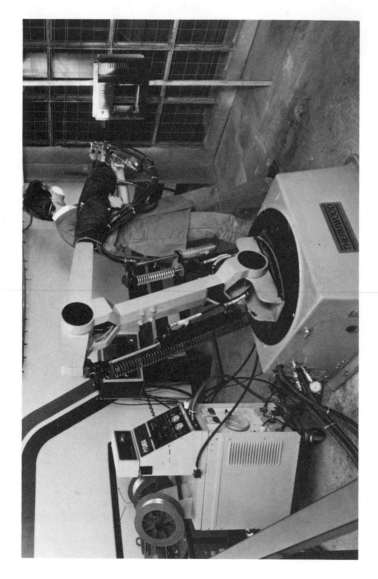

Lead through teach programming of a hydraulic, jointed arm, servo, spray painting robot *(Photo courtesy Thermwood Corporation)*

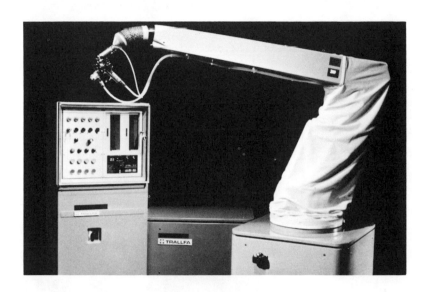

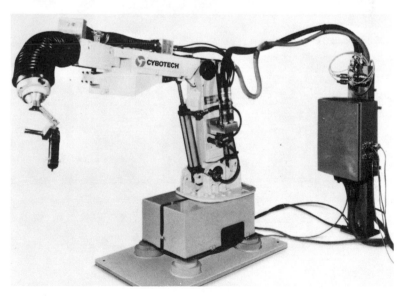

Hydraulic, jointed arm, servo spray painting robots *(Top photo courtesy the Devilbiss Company; Bottom photo courtesy Cybotech Corporation)*

Electric, jointed arm, servo robot with production set up for arc welding *(Photo courtesy Hobart Brothers Company, Troy, Ohio)*

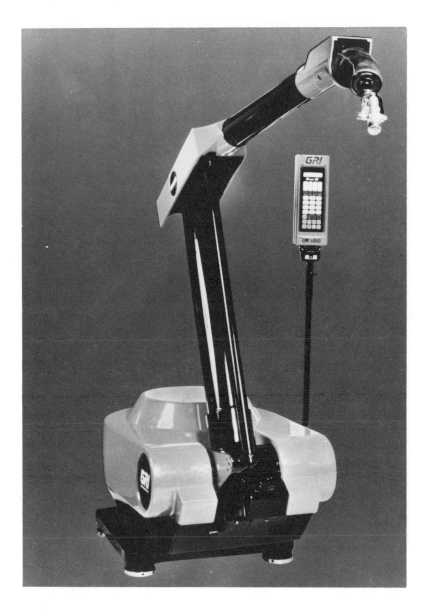

Hydraulic, jointed arm, servo robot for spray painting *(Photo courtesy Graco Robotics, Inc.)*

Hydraulic, jointed arm, servo robot shown during lead through teach programming for material transfer *(Photo courtesy Thermwood Corporation)*

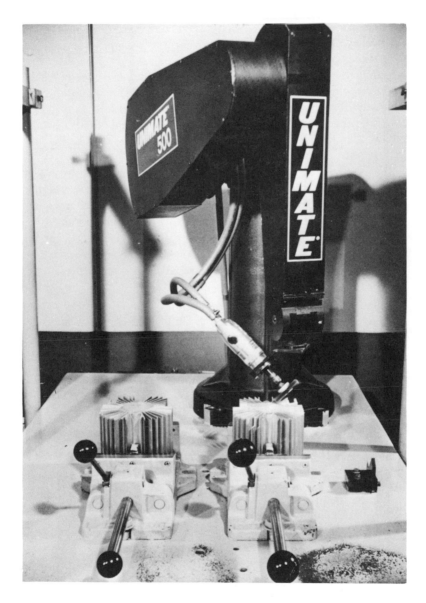

Electric, jointed arm, servo robot for assembly and light duty processing *(Photo courtesy Unimation, Inc.)*

One example of this is an insulation coating sprayed on nose cones of missiles. A spray painting robot is utilized to apply the material even though the production rate is less than one part per week. The reason is that the coating which is placed on the nose cone is highly toxic, and the operation has to be conducted in a special, completely enclosed room. In this instance, the cost and installation of the industrial robot had to be weighed against other possible methods of completing the task while maintaining operator safety, such as placing the operator in a completely enclosed "space suit."

It is obvious that too low a production level can make the use of industrial robots impractical. What is not obvious, however, is that once production levels become very high, other high speed fixed automation equipment may prove superior to industrial robots.

If millions of a part are going to be produced with little or no change, a simple, fixed transfer device or specialized machine can be built at or below the cost of an industrial robot. That special machine can out-perform the robot.

Making a decision between fixed automation and industrial robots can prove very difficult. I can offer no simple rule of thumb and suggest that several areas be given consideration.

I have already touched on the first, that is, a consideration of the amount of production necessary from the equipment. A process which runs continuously day in and day out, unchanged, is a very likely candidate for fixed, special purpose machinery. Even certain processes which run relatively large quantities with infrequent, minor changes may be candidates for fixed automation, although a higher level of flexibility is now necessary. As the level of flexibility necessary to produce the product increases, industrial robots begin to make sense.

Another consideration is the life of the product. One characteristic of fixed automation is that major product changes make the production equipment obsolete. This, of course, is not the case with industrial robots. If a product will be manufactured unchanged for many years, the cost of the necessary fixed automation can be easily amortized over the life of the product. Retiring the equipment at the end of the product cycle is not an economic burden. Products, however, which might change each year or so and have an unknown overall life cycle might lend themselves to the use of robots. If a product line change eliminates the need for the robot, it can be simply reprogrammed and moved to a different application for a different product. In this way certain safety factors are built into the cost of a product's tooling, providing more flexibility in reacting to changes in the marketplace.

In order to determine when an industrial robot makes sense in high volume production, you might ask yourself whether you would consider having a person perform the task. In those instances in which a person would be the first choice, you have found a good application for an industrial robot. In those instances in which a piece of fixed equipment could easily pay for itself by replacing the manual labor, it probably should. The gray area between these two extremes needs to be analyzed on an individual basis before making any decisions.

What Is the Ultimate Robot Utilization?

In using industrial robots there is a simple rule: "If you only need one, you probably don't need any."

As you can tell by now, there is an effort required in using industrial robots. There are costs involved in getting through the learning curve, establishing the necessary programming capability, and training operators. A maintenance schedule, training, and maintenance part inventory need to be established. If this entire effort is undertaken for a single robot, it will quickly eliminate any savings derived from the robot. If there is a commitment to a continuing program, the costs and efforts make sense.

By this I am certainly not recommending that a massive program be established at the start or that a multitude of machines be installed initially. A slow, careful approach to robotics is recommended, beginning with a single installation. Once that first evaluation installation has proven itself and the expected benefits occur, there should be a real commitment toward building on that success by installing additional machines.

I do not want to imply that the follow up installations need to be identical to the initial installation. The knowledge gathered from a successful robot installation can be applied to a variety of different applications within a plant. The general type and complexity of the robot initially installed determines the level of capability that you have developed. Installing a simple air powered, pick and place robot will not give you the background needed for operating a bank of five servo controlled, continuous path spray painting robots. If you need five spray painting robots, do your development on a single installation with one robot first. Once you are comfortable with the machine and its capabilities, the other four can then be installed as the follow up.

Also try to keep the robots within your facility as common as possible. Customizing robots to the extent that each installation is a highly specialized robot machine provides little flexibility over fixed automation. The universal automation appeal of robots—that they can be moved from job to job as production changes warrant—disappears when the robots are no longer universal. By maintaining some universality, you can simplify machines, training requirements, maintenance, spare parts, and operator training. This approach may require that in some instances robots will be performing functions which might be better suited to a slightly lower cost but radically different machine. If there is a substantial difference between the lower cost machine and the "standard machine," pricing must take precedence. The cost of underutilizing a standard robot will show up at the initial purchase price; the cost of the lower priced, lower capability machine may appear later in the form of increased training and maintenance costs and decreased overall flexibility.

Again, unless there is reasonable expectation and established need for the installation of a number of industrial robots, you should not be using any at all.

Can It Be Justified?

This question probably should have been asked first in this chapter. It is the key in trying to determine if a particular application is suited for industrial robots. Until now we have concentrated on whether or not the particular task was practical or feasible. The real question is whether, from a dollars and cents standpoint, you even want to install an industrial robot.

Caught up in the excitement of a new technology, many firms are installing industrial robots without economic justification. They are installing robots just for the sake of having robots. The fact is, however, that sooner or later robots are going to have to pay their way in a free market system. It is therefore best to have them pay their way right from the beginning. Industrial robots should be cost justified on the same basis that any other capital expenditure is justified. This justification needs to be very thorough. To assist you in determining the economies of an industrial robot, Chapter 8 of this book contains a detailed analysis of an industrial robot installation with some additional thoughts on real world economics.

It is sufficient to say that a robot has to make you money and has to provide an acceptable return on investment. If a robot installation cannot pass this initial criterion, then the remainder of the discussion about its implementation and installation is purely academic.

Chapter Six
DOLLARS AND SENSE

It's now time to enter the uncharted waters of robot applications. Robots are the same as, and yet different from, other pieces of automated equipment. They will entail changes in methods and changes in thinking. Therefore, they will require, at least in the beginning, a higher level of concern, analysis, and effort than more familiar pieces of industrial equipment. As a guide to your first industrial robot installation, I provide the following five commandments. A transgression against any of these rules invites the wrath of the gods and sets you up for innumerable problems. I assure you that the efforts required to follow the following five rules carefully and completely will be considerably less than the efforts required to solve the multitude of problems that will result if you don't.

These five cardinal rules of robot installations are very general, but they are intended to point you in the right direction and steer you away from those areas that can cause additional problems in initial installations.

It should be understood that I am commenting on the initial installations only, and these rules may not apply with the same vigor to subsequent installations once a body of knowledge and experience has been gained.

Rule Number 1—Think Simple

The potential applications for robots can be ranked in a number of different ways. Applications can be listed based on the potential savings available. Using this method, the installation providing the greatest overall economic benefits is listed first, and the others follow in descending order of economic importance. Potential installations also can be ranked by complexity. In ranking potential applications in this manner, the simplest and easiest overall application is ranked first with others listed in increasing levels of operational complexity and installation difficulty. Applications can be listed by a number of other criteria including impact on labor, improvements in capacity, or reductions in undesirable tasks. The task of determining which job should be approached first takes on a rather complex character when the decision-maker has this many ways of analyzing the situation. Seldom, if ever, will the same task rank at or near the top in all of the different ranking systems. This means that it could become necessary to determine which of the various criteria are most important for the initial robot installation.

The rule of thumb here is to "think simple." The initial application should be the easiest to install and engineer, while also providing a reasonable level of economic benefits. In making this choice it must be understood that you may be passing up many applications with higher overall return on investment. If a commitment to robots has been made, an implementation plan should include the installation of more than a single unit. With this in mind, those higher payback, but obviously more complex, applications can be handled in the future by a more confident and more experienced team that has proven successes to its credit in the simpler applications. The odds of success for the entire robot program are enhanced dramatically by being a bit more cautious at the front end. A successful robot installation, even in a simple application, will quiet the skeptics, convert the doubters, quell most fears of labor and management people, and set the stage for substantial future successes.

Should a more complex but higher payback project receive initial attention, the odds of success are reduced. There are two reasons for this. First, the applications job is more complex. This is normally not a problem for an experienced organization. The second reason, however, is that during that first project your organization is at the lowest point in its knowledge and practical experience with robots. Burdening this inexperienced organization with complex initial installations is asking for trouble.

I am not discouraging complex, bold plans for the use of robots. These bold, integrated, multi-robot applications are the real future of robotics and are also where the major costs savings can be realized. It must, however, be understood that no organization can implement these plans overnight. A normal learning curve exists in the application of industrial robots as in all other new business endeavors. An understanding of this important fact will lead the astute organization on a planned, careful, and conservative initial journey into the world of robotics. This careful journey through the learning curve will provide ample rewards through future smooth and successful robot installations.

Rule Number 2—Be Thorough

Once the initial installation for an industrial robot has been selected, the task of applying the industrial robot begins. The greatest dangers you face now are surprises. The success or failure of your initial installation will not depend as much on the quality of those things you do as on the quantity of those things you don't do. It is the unexpected surprise that trips you up: the production technique that everyone is aware of but does not appear on the written production standard; the production modifications necessary on 2% of the orders run; the unreported, uncorrected problem on machine number one to which the human operator adapts. It is necessary to develop a complete, intimate knowledge of the area, the process, the variations, and the production where the robot will work.

The first step in accomplishing this includes getting production employees and supervisors involved with the application. This involvement provides you with two major benefits.

First, no engineer, no manager, no time study person is more familiar with the task to be performed than the person performing it. His or her input is the most comprehensive guide to the requirements of that particular task. Of course, approaching the person now performing the task requires answering his or her questions concerning how the contemplated changes will affect that person directly. Providing this information in detail at an early stage is another recommended procedure and will be covered in a later section.

The second reason for getting both the production employees and supervision involved in the planning stages of the installation is to enlist their support. They have a stake in the successful installation of the robot. By their actions or inactions they can control the ease or difficulty with which a robot is implemented. Getting these employees involved early and making them feel like a part of the installation will also reduce their tendency to blame the robot for every problem that occurs throughout the installation.

Being thorough involves more than doing just a super job in designing, engineering, and installing the robot system. It means handling the various tasks necessary for long-term profitable use of industrial robots. The installation must be maintained. Regardless of manufacturers' claims, robots are machines. They will break down, and according to our friend Murphy, they will break down at the worst possible time. Make certain that your maintenance personnel have had the proper comprehensive training necessary to maintain the installation. This training cannot be restricted to one individual, especially if you run a multi-shift operation. Maintenance coverage and backup are necessary on all shifts during which the robot will be operating.

This maintenance training will do little good if the proper spare parts are not available to your maintenance personnel. Most robot manufacturers have recommended lists of spare parts which provide a guide to setting up your maintenance operations.

The time will come, however, when some component fails which you do not have in your spare parts inventory. Some provisions must be made for handling that situation. If production can be interrupted while repair parts are obtained, the solution is simple. Be certain you understand the repair parts policy and part availability of the supplier you have chosen to supply your robot. If a production interruption for that period of time is unacceptable, other alternatives should be considered, whether it be returning to a manual operation or keeping a whole spare robot on hand for such emergencies. Don't wait until this problem arises to develop a solution.

Robots require something many other machines do not require—programming. If you operate any other NC or computer controlled machine, you already understand this. If you do not, it becomes a new consideration. Again, someone must be responsible for developing the various programs that will be used. It is a good idea to have more than one individual thoroughly familiar with programming. There should be at least one person on each operating shift who can make minor adjustments to the robot's program, unless the trained programmer doesn't particularly mind waking up in the middle of the night and coming into the plant to make these minor corrections.

The details that must be considered in a robot installation are numerous. Rather than try to cover all of them in this section, I have included them in a robot installation manual developed to guide you through the planning, engineering, and installation phases of an industrial robot program. This manual is reproduced in Appendix I of this book. It consists of a number of questions, each of which should be answered in detail. Instructions on the use of the manual are included in Appendix I.

I cannot overemphasize the importance of being thorough. Probably the largest number of robot installations which have been unsuccessful due to engineering deficiencies could have been successful had the responsible individuals developed the thorough understanding recommended here.

Rule Number 3—Be Reasonable

Industrial robots contain absolutely no magic. To avoid disappointment, be reasonable in your expectations of what the robot will and will not do. An industrial robot will not make order out of chaos. With the exception of a few research oriented machines equipped with sight or other sensing devices, an industrial robot will require a more structured, more consistent environment to work in than its human counterpart.

Be reasonable in your expectations for installation and debug time. Regardless of the level of planning, it is virtually impossible to install the robotic system and have it function perfectly the first time. There are always details that must be resolved. These problems, however, can be handled one at a time until they are solved. On the other hand, the pressures caused by allowing insufficient time to solve the inevitable start-up problems unnecessarily complicate the installation. Make certain that contingencies have been made to handle lost production during this debug phase.

Consider the robot's capabilities. Make certain that the consistency of part placement with which the robot must work is sufficient for the particular robot and hand configuration chosen. Make certain the weight carrying capacity and speed of the robot are sufficient to perform the necessary tasks in the time allotted.

The weight, speed, accuracy, reach, and other physical characteristics of the machine deserve special mention here. One of the most unreasonable things that can happen in robot installations is to design the installation to use every last ounce of capacity in one area or another. By doing this and providing little or no safety factor, many of the simplest solutions to start-up and debug problems are no longer available to you. When something unexpected occurs during the start-up phase, it is wise to have some reserve capability in each of the physical characteristics of the robot to assist in overcoming the problems. One of the quickest ways to discourage enthusiasm for robots is to allow the physical limitations of the robot to become the bottleneck that prevents the successful automation of a production task.

Along the same line, be careful that you don't believe everything that the robot manufacturer tells you, including the specifications printed on the manufacturer's literature. While I am not contending that any manufacturer would try to deceive customers deliberately about the capability of their machinery, it must be understood that product literature and specifications portray the manufacturer's products in the best possible light. The equipment may not be able to match every specification to the letter, so if a particular specification must be met for your application, be certain to test the equipment before proceeding.

One of the large, multinational oil concerns maintains a laboratory and research center in The Hague, Netherlands. Any vendor or manufacturer that would like to supply product to this company anywhere in the world must first obtain approval. The approval process is simple. The manufacturer's product literature and product are submitted together. The testing consists of simply verifying that the product meets the specifications listed in the product literature. If it meets the specifications, it becomes an approved product. If it does not meet the specifications, it is not approved. It is surprising how many products on the market do not get approved.

This can be traced to the fact that the testing procedures used by the manufacturer may be quite different from those used by the customer. The specifications for a $100,000 Rolls Royce say that it will go 140 m.p.h. and that it will also turn in a 30' diameter circle. However, if you go down the road at 140 m.p.h. and try to make a 30' diameter turn, you might discover that it does not meet specifications. It's all in how the test to determine the specifications is conducted.

Along the same line, it is not realistic to depend totally on the robot manufacturer for the applications engineering needed to implement the robot. The manufacturer cannot understand your production situation as well as you and your people do.

Be reasonable in your expectations of support from the robot manufacturer. Those manufacturers that maintain large, well-staffed development and applications engineering groups can handle your applications work better than those small firms that rely on a few individuals for all of their tasks. The applications engineering groups, however, cannot be expected to do detailed applications engineering work as part of the purchase price of the robot. In fact, most of the manufacturers that maintain these groups operate them as separate profit centers, and you should expect to pay for their efforts. After all, you will pay for this work whether your own engineering people do the job, whether you hire an independent consultant, or whether you utilize the applications engineering department of a major robot manufacturer.

Once engineered, developed, and started, a robot installation will not simply continue running unchanged month after month. Things change, sometimes gradually, sometimes quickly. It should be expected that adjustments and changes will be needed in the robot program on a continuing basis. Expecting less than this will lead to disappointment in the robot operation.

Because I have been focusing on the industrial robot and its engineering and applications, you may have grown a bit complacent about the equipment, tools, and supporting devices in your installation. Don't assume that the only problems that will occur will be robot problems. Problems with tools, fixtures, supporting equipment, and the like will also occur. Be certain to give them the same overall level of attention that you give the robot.

Make certain that your expectations of the cost of the installation are realistic. In most applications, the robot cost is only one part of an overall package. As a general rule of thumb, installations utilizing low-cost pick and place robots will run three to four times the robot cost. Medium technology installations will cost two to three times the robot cost, with those installations utilizing the high technology, high capability robots costing between one and one half to two times the robot cost.

Being reasonable means using a good deal of common sense in choosing and implementing your robot program. Be realistic in your expectations and your planning, and the successes you desire will result.

Rule Number 4—Be Honest

This rule has two parts: Be honest to yourself and be honest to others.

I'll begin with being honest to yourself. An honest appraisal of the company in which you work is valuable in trying to implement new technology. For example, a lack of engineering or technical capability is not nearly as big a problem as not recognizing the deficiency at all. Technical advice and engineering assistance are available from a variety of sources. However, this assistance will not be available unless you seek it, and you will not seek it unless you believe you have a genuine need for it—hence, the need for an honest appraisal of your strengths and weaknesses.

The most important application of this rule is to be open and honest with others in the plant, from the workers who will be operating with and around the robot to top management personnel who have a keen interest in the robot installation. It is vital that the production people and production supervision be appraised constantly and honestly of the progress, problems and implications of the robot installation. These people are genuinely interested in the proposed installation. The installation will directly affect the way they perform their jobs as well as their job security and their future. The feeling that the robot will cause pronounced change is experienced by not only those workers being moved because of the robot, but also by those who now must substitute robots for those people that they supervise. It is important that these individuals feel that the changes that will occur will be beneficial and are not detrimental to them individually. Their help and cooperation will not come unless these people feel secure, and that security only comes from an open, honest approach by management. Any attempts to whitewash, cover up or present the robot in other than completely honest terms will be quickly detected and will lead to suspicion, doubt, and resistance. Obviously, for this approach to work, the robot must prove beneficial to the production organization. This, however, is normally the case, since displaced workers can be retrained or transferred to better, more interesting jobs.

This honest appraisal of the robot installation should extend to your community relations. Since robots are a very newsworthy item today, you can be certain that local news media may very well have an interest in your proposed installation. To assure a fair and accurate reporting of the robot's implications, some effort is necessary to communicate to these people the benefits of automation not only for your company and the country in general but also for the individual workers within your plant. It is only in this manner that industrial robots will be portrayed as beneficial helpers rather than mechanical villains causing unemployment.

Once robots are installed and are operating successfully, the need for this careful effort can possibly diminish somewhat. In the early stages of the first robot application, at least in today's world, an honest, straightforward effort in disseminating information is vital.

Rule Number 5—Be Careful

A large portion of care and caution is called for in a robot installation.

Robots have an interesting operating characteristic. If you watch a robot performing a task normally done by a person, you see a machine moving very smoothly and performing a task just like a person. After watching the machine continuously perform its function day in and day out with the smoothness and dexterity of a skillful human, it is very easy to forget that the robot is in fact a machine which can malfunction. The real problem here is that industrial robots are many, many times stronger than a human being and much faster. They are capable of causing many serious injuries that a person could not possibly inflict. It is necessary, therefore, to protect your people from the robot. Broken hydraulic lines, malfunctioning control systems, stuck valves, and the like are possible on every machine manufactured today. These malfunctions can move the arm swiftly and uncontrollably to any position within the operating envelope. People must be physically restrained from being in the operating envelope any time the robot is in operation. Warnings about the danger inherent with a robot malfunction are quickly forgotten as employees watch the machine moving smoothly and controllably day in and day out. The only way to protect the employees properly from the robot is with a very safety-conscious installation design. I, for one, do not trust electronic safety controls such as electric eyes, light curtains, and the like. These can malfunction with serious consequences. The only truly safe way to operate a robot is with a physical barrier such as a fence, a gate, a guardrail, or the like, between people and robots. Access doors which allow maintenance crews and the like within the robot operating envelope should be interconnected in such a way that opening the doors turns off the robot. Any safety procedure less positive than this is inviting disaster.

Those persons working on or around the robot should be fully appraised of the safety procedures and dangers inherent with an industrial robot. An explanation of possible malfunctions of the robot should be provided along with details concerning the safety precautions built into the system. Nothing will turn your employees against robots as quickly as having one of them injured by a runaway machine.

Also protect the robot from your people. Some facilities have more problems with vandalism than others. While it is not a common problem, people have attacked robots. While there are no firm rules on how to protect your robots from your people, normal common sense should be used. Keep the programming devices in a safe place, and keep your control panels locked so that detrimental changes to the system are not easy to accomplish. If designed properly, the same safety gate that protects the workers from the robots can also protect the robots from the workers. Make certain that errors caused by misplaced parts, limit switches tripped at the wrong time, chewing gum on the conveyors, and the like can be detected and recognized as minor attempts at sabotage rather than as failings of the installed system. Make certain that there is some way to detect malfunctions in the robot installation. Placing the robot in a room by itself and leaving it unattended to perform its task for long periods of time invite problems. A malfunctioning robot is capable of producing large quantities of scrap, just as a well-operating robot is capable of producing large quantities of good quality product. Installations must therefore be tended in some way, either by nearby workers or computer monitors. It is seldom that a robot in operation is so isolated that it is not within the view of a responsible individual that can keep an eye out for obvious malfunctions.

Be careful with your robot programs. Most programs developed for an application have a great deal of value if you consider the number of planning, management, engineering, and supervision hours which have gone into its development and debugging. Make certain you have adequate records of your program. Most robot systems provide some method of storing the various programs on cassettes, punch tape, magnetic discs, or other media. It is common sense to maintain a library copy as well as the copy used to load and operate the robot. Should a malfunction occur which destroys one copy, a backup is then available. Make certain that the backup or library copy is kept current and reflects all of the improvements and changes which have occurred to the operation. It is also a good idea to maintain a written record of a program and program sequence, should it ever be necessary to develop the program from the beginning.

The reason for a cautious approach in maintaining programs is that a number of failures can and do occur on computers which can completely destroy any data stored in the system. Even so-called non-volatile memories (those that do not lose their information when power is turned off) have certain failure modes which can completely scramble or destroy all of the information present. Without a backup it would be easy to lose weeks or months of effort in a matter of a few seconds. In few other areas in business can so much be lost so quickly with no hope of retrieval as in having poorly documented, poorly backed computer programs.

As in any other endeavor, the first robot installation is the most difficult. More bases must be covered more carefully. Because robots conjure up colorful visions and deep-seated fears, a generally greater effort must be expended in the initial robot installations than is required with most other pieces of industrial equipment. Hundreds of firms just like yours, both here and abroad, have successfully installed industrial robots and are benefiting daily from their operation. Once this initial hurdle is overcome, which does require a commitment and a real effort, successive installations will prove much easier and possibly more profitable.

Chapter Seven
THE AUTOMATIC FACTORY

With the advent of computer controlled machinery and industrial robots and with the aid of ever increasing sophistication in management computers, a phenomenon called the automatic factory is emerging. This term, sometimes referred to as a flexible manufacturing system, or FMS, refers to a production arrangement whereby computers run the factory. In its extreme form it is an unmanned factory. In these systems a hierarchy is set up in which various pieces of production equipment, robots, material handling equipment, and the like are each controlled by its own computer. Each of these computers, in turn, is controlled by supervisory computers which schedule the work. These supervisory computers then are controlled by the central management computer which takes its instructions from the company management. With this arrangement, production runs of only a few parts can be produced economically and automatically. Setup, material handling, production, quality control, packaging, and inventory of the products are handled with little or no human intervention. In even more sophisticated applications, design computers assist the product designers in developing the product and programming the various robots and production machinery to produce that product.

In a more practical vein, the automatic factory generally is designed to manufacture a related family of parts. As an example, a facility set up to produce electric motors might be flexible enough to produce a wide variety of electric motors which differ in size, power, shape, and the like. This same facility, however, would not be capable of producing lawn mowers or garden hoses. Once this is understood, the feasibility of automating the production of a family of products becomes more realistic.

If you examine most products manufactured today, you will find you can automate 80 to 95% of their production using fairly conventional technology, given enough time and money. You will also discover that the final 5 to 15% requires judgment, dexterity, or adaptability that machines simply do not have today.

Faced with this, there are three possible solutions. First, and probably easiest, is to inject people into the system to handle those tasks not easily handled by machines. The engineering purists who desire a totally automatic factory shudder at this thought and consider it a "cop-out."

Your second choice is to simply wait until someone, somewhere, develops the necessary sensory perceptions for your robots and machines so that totally automating the facility becomes practical. Many individuals and companies must believe that this is a viable approach, since a great deal of time, effort and money is being devoted to the development of special sensor technology such as sight, tactile sensing, and the like to handle these varied and unique situations. Based on the resources being devoted towards these ends, it is very likely that many of the techniques and much of the technology being developed will become practical and cost effective. Those people developing systems today and those stout-hearted individuals attempting to utilize many of these systems will be looked on as the true pioneers of the robot industry. Please remember, however, that it is very easy to tell who the pioneers are: they are the people with the arrows in their backs.

The third choice in solving your problems involves redesigning the product so that the troublesome 5 to 15% no longer exists. This means redesigning a product so that the manufacturing requirements match the manufacturing capabilities of your automatic factory. While this sounds like the simplest and most straightforward way of handling the problem, it certainly has its share of pitfalls. The only truly efficient way of accomplishing this goal is by using a computer aided design system totally familiar with the capabilities of the automatic factory.

Unfortunately, the systems that are available today are very expensive, and none that I am aware of is fully integrated into this type of automatic production system. This approach, however, does hold a great deal of promise for the future. With the advances which are occurring in computer capabilities, I will predict that this is the approach which eventually will win out.

Of all the approaches mentioned to developing an automatic factory, I know of none which uses the high level of sensory perception suggested. A good deal of work is going on to develop a computer aided design and automatic manufacturing combination which would implement the last approach mentioned. There are quite a few installations, some quite sophisticated, which use the first approach suggested. By integrating people and automated machines under the control of scheduling computers, some very sophisticated and cost effective operations presently are manufacturing product.

Development of an automatic manufacturing system similar to the types described here should only be attempted by a team of highly competent technical specialists, experienced in automation and very, very well funded. Even then, there is a good chance of failure. I cannot emphasize too strongly the difficulties encountered in attempting to implement this type of system. Such a large part of this type of system must be custom built today that the chances of any reasonable financial payback become very slim. Given some time, with the efforts of our pioneers—and some pioneers stumble across gold mines—these types of systems will become commercially available, dependable, reliable, and predictable. If you are in business to make money, that's the time to get one.

Right now, however, a lot of people are talking about automatic factories and flexible manufacturing systems. So I'd like to examine the various aspects and components associated with an automatic factory.

CAD/CAM

CAD/CAM stands for "computer aided design/computer aided manufacturing." This broad generic set of terms is used to denote those areas in which computers are used to aid in either the design and/or manufacture of a product. Computers used in the manufacture include those systems used in the NC control systems within industrial robots, material handling equipment, and the like. This section, however, will concentrate on the use of computers to aid the product designer.

An entire industry is developing around using computers and graphics terminals to design and develop products. As with most computer equipment, systems vary widely in their capability, sophistication, and costs.

A computer aided design system begins with having the designer put a drawing or design into the system. This is accomplished in a variety of ways. One method involves specifying the beginning and ending of each line segment using some convention. A variety of specialized equipment is available for this function. This specialized equipment includes special drafting tables, light pens which can indicate positions on CRTs, and a specialized key function on CRTs which allows geometric shapes to be input.

Some of the more sophisticated computer aided design systems actually allow the input of the physical properties of the material from which the part will be made. It is then possible to calculate various stresses that the product will encounter during its manufacture and use. The end result of virtually every computer aided design system is a data base describing the geometric characteristics of the product.

Machining the part which has now been designed requires a human programmer. Using a design system called Apt, the programmer develops the path of the center line of the cutting tool with respect to the part. This geometric information is called the center line information, or CL information. The format of this particular information is somewhat standardized today, unlike the geometric outputs which can vary widely from manufacturer to manufacturer.

The CL information now must be processed to provide an instruction tape, or NC tape, for the particular machine tool which will machine the product. This is done with a computer software program called a post processor. Since the format and instructions of various machine tools are different, each machine tool type requires a specific post processor.

Past practice has dictated that from the post processor a punched paper or mylar tape is developed. This punched tape contains holes approximately 1/16" in diameter whose pattern carries information understood by the various computers. This punched tape must be carried physically to the particular machine tool for which it was prepared, loaded, and then used to manufacture the part. Each machine tool then had its own tape reader.

Recently there seems to be a trend toward direct numerical control or "DNC," in which the program information for all of the machine tools within a factory is contained in a main computer and storage system and is transmitted as requested to a particular tool. Each tool is wired to the main control bank. This interconnecting system provides the rudimentary network necessary to begin building an automatic factory.

There is a great deal of effort being conducted to eliminate the Apt programmer and allow a computer to generate the center line information automatically. To date I am not aware of any systems capable of doing this in more than two axes, thus relegating their use to two dimensional parts.

Future efforts seem to be directed toward making the flow from a computer aided design system to the actual manufacturing floor more direct with less human intervention. A number of machine tool manufacturers presently offer commercial systems called flexible machining systems. These systems utilize fixture carts which move machining fixtures and part blanks from machine to machine. These carts normally track tape, wire, or other devices either on the floor or in the floor. They deliver a properly-fixtured part to a machine tool and load the fixture and part. Then, after machining, the carts again accept the fixture and partially machined part for transport to the next station. Such systems, when combined with sophisticated machine tools capable of accessing a large number of different cutting tools, hold the promise of dramatically reducing the cost of small bath machine parts.

Automatic Storage and Retrieval

In order for a completely automatic factory to function, it will be necessary to integrate automatic warehousing systems with the various assembly and machining operations controlled through CAD/

CAM. A large variety of rather sophisticated automatic warehousing systems are commercially available today. Some of these systems utilize large special purpose buildings with specially built racks, high lift fork trucks, and pallets. The driverless fork trucks and conveyor systems all are computer controlled and can store or retrieve stock very quickly. Other systems use driverless fork trucks which again follow tape, wire, or the like located on the floor. I have even seen automatic fork trucks with infra-red "eyes" at the tips of the forks that indicate to the control computer when the forks are properly positioned to pick up a load.

The automatic factory will require the integration of automatic warehousing and retrieval systems, flexible manufacturing systems, robotics, and hierarchal computer control. Unfortunately, today no single company operates in all of these areas. Each group specializes in one part of the whole, and in many cases these companies so jealously guard their technology that it is nearly impossible to obtain the cooperation necessary to integrate their systems with other manufacturers' machinery. In Japan efforts toward this end have been somewhat successful, rather dramatically reducing the number of employees in a typical plant. These efforts have, however, required massive government financial assistance and a certain level of governmental arm twisting, neither of which exists in the United States today.

I have included this rather sketchy overview of the automatic factory to provide you with at least a basic knowledge that such things exist. Although people are talking about and working toward the automatic factory, a true realization of its potential in all but a few very well capitalized industries is a number of years away. Most companies within most industries are finding it difficult enough to cope with using robots to handle simple tasks. At times, there's a certain temptation, especially on the part of farsighted top executives, to strive for the ultimate in this new area. Taking that route today is financially dangerous and filled with substantial technological risks. In the next few years a maturing will occur that will make more aggressive integrated factories more practical. This maturing will involve technological improvements, increased experience on the part of users, and increased confidence on the part of those associated with the programs.

Chapter Eight
HUMAN RELATIONS

I recently attended *The Empire Strikes Back*, the sequel to the unbelievably successful motion picture *Star Wars*. This film brings to life a liberal sprinkling of technical options and fantasies. The effects achieved in the movie were quite phenomenal. The film makes extensive use of machines which we have come to know as robots. An interesting contrast is developed throughout the show. The popular, friendly, almost humanlike robots called R2D2 and C3PO, well-known to the American public, provide the capable help to their human masters that we feel robots are destined to do. These roles portray all the most desirable traits of industrial robots. However, there was also another type of robot shown. These appeared as gigantic, metal, four-legged animals thousands of feet tall that brought death and destruction to all in their path. As seen from this vantage point, robots personify all that is undesirable and evil with advancing technology.

The people within your organization and within your community will recall these visions when you announce that you are about to install industrial robots within your facility. Whether their concept of the robots you are going to use is that of helpful, positive droids like R2D2 or the negative, destructive robots personified by the war machines will depend entirely on you. If you take the time and put forth the effort to slowly, carefully, and accurately inform all the interested parties about the benefits of using industrial robots, the results are obvious. If you are careful about the installation so that it benefits each of the groups associated with it, the machines will be viewed as beneficial. If, on the other hand, you simply ignore the questions, trying to force the machine in with no consideration of individual attitudes or opinions, it would be normal and natural for the robots to be viewed as a threat. Visions of destructive mechanical creatures spreading unhappiness and misery will be applied to the machines you are intending to implement.

The efforts required to do a good job of properly informing all concerned parties are not that great. The minimal exertion it takes to properly introduce the new machines can easily be ignored in the haste and excitement of trying to install an industrial robot. These simple efforts, however, must not be omitted, as they are quite vital to the long-term success of robot implementation.

I have purposely called this chapter human relations and not labor relations because more than just direct labor must be considered. In addition to considering your employees' views, if you have a union, you must take into account the union's position on industrial robots. You must also consider the reaction of your production supervision, top management, and the community in which you reside. Each of these groups views the industrial robot differently, and each can have a measurable effect on your installation.

Production Employees

For the moment put yourself in place of one of your hourly paid production employees. As a production employee, the first thing you notice is that your knowledge of what is happening within the company is extremely limited. That knowledge is controlled almost entirely by those in middle and upper management. The information you do get is quite valuable to you. If this is a typical company, it is also quite scarce. Certainly the day-to-day details concerning facility operations are totally unknown to you.

Looking back over the last several years, you also find you cannot depend on what top management tells you. As an example, when that new automatic machine came in and replaced three workers, management assured you that no one would be laid off as a result. A month and a half later three employees in Department Three were laid off, and you are certain it was a direct result of the automatic installation. Your manager knows that the layoff was in no way connected with the new installation, but was controlled by other economic factors. The fact that the layoff of three people was exactly equal to the number of people replaced by the automatic machine was a coincidence. Notwithstanding, your manager's knowledge of the situation and your knowledge of the situation as an hourly employee are quite different.

Misunderstandings in which hourly employees see and believe one thing while top management understands and believes something almost totally opposite are quite common in industry today. These misunderstandings are the result of a lack of rapport or a lack of communication which breeds a level of distrust that exists to one extent or another in almost every industrial operation today. The truth is that it would be virtually impossible to keep the production workers so thoroughly informed that they would understand the various decisions made on a day-to-day basis. At the same time, the fact that this communication problem and potential for misunderstanding exist must be realized.

As an hourly paid production worker, another feeling which you have is that production supervision and management seriously underestimate your intelligence. They don't seem to realize your full potential, and they treat you very much like you treat your children. You resent this, especially when they tell you things you know are wrong or misleading and then expect you to believe them. In short, many managers and many companies are guilty of not recognizing the inherent intelligence and common sense of their production workers.

This little exercise points out one very important fact. The hourly paid production workers within your facility view your business and production efforts quite differently than you do. Their view of what is going on around them and within the company is determined by what they are told by management and what they see occurring each day. Any time there is a conflict between what they are told and what they perceive, they will naturally believe what they see.

Honesty is the real key in dealing with hourly paid employees concerning robots or other types of automation. An early, honest, open communication to these workers, presenting all the facts, is an important part of a successful installation. In order to gain the support and backing you require, the robot installation must be beneficial to the hourly paid employees and must be presented from the start in such a way that it also appears beneficial.

The hierarchy of needs for employees has been discussed in depth by many authors. It begins with the basic need for food and shelter and moves on to higher needs, including interesting, challenging work, recognition, etc. Make certain your installation does not try to substitute a small, positive move in one of the higher, less important needs, while providing a reduced level of individual security in one of the most basic, and hence more important, needs.

A cardinal rule of installing automation is that it must not cost anyone his or her job. An overall reduction in work force because of the automation is generally acceptable provided such a reduction occurs through natural attrition and not through forced reduction or layoff. This approach may prove more expensive in the short run when compared to immediately eliminating those workers whose jobs have been automated. The long-term rewards for the additional expenditures can be substantial. Make certain that those workers whose jobs are changed move to jobs which they consider more desirable. These jobs can be more desirable because they are less fatiguing, higher paid, more challenging, and the like.

On the other hand, allowing a robot installation to reduce the desirability of people jobs is almost as bad as allowing the automation process to directly cost workers their jobs. Besides the insecurity felt by co-workers as they wonder when they will be taking their step down, those workers who have been moved to the less desirable jobs have ample opportunity to complain vehemently about their plight. This type of dissent understandably can be directed toward the industrial robot which caused the problem in the first place. This reaction can take the form of direct sabotage of the equipment. However, usually the reaction is not an out-and-out destructive act, but a slow, calculated plan to cause the robot to fail. In any production facility a joint effort by hourly paid production employees to cause an installation to fail will succeed. These people know in fine detail all those things that can make the robots' task both possible and impossible. Sabotage in this manner is virtually impossible to detect or prove, but it quickly will remove the potential inherent in the use of industrial robots.

Labor relations should not be left to chance. A little effort in properly planning the robot installation so that it harmonizes with the production workers' concerns, together with a program to properly introduce these workers to the benefits they will receive, will dramatically increase the chances of successful implementation.

Production Supervision

In examining the reactions of foremen or first line production supervision to a proposed robot installation, keep in mind several facts about their jobs. First and foremost, these people are in the front lines and under a great deal of immediate pressure. Their job is to get out production with the tools, the methods, and the personnel available. If production doesn't occur in the desired manner, explanations will not suffice. Production successes yesterday or last week don't count today.

Most foremen that have been on a job for any time at all have seen what good intentions, poorly executed, can do. They have seen well-intentioned moves by engineering or top management which were intended to increase production or save money have just the opposite effect. With a belief that their present production systems are fine and adequate, they view changes with suspicion.

The successful foreman understands in minute detail all aspects of the jobs he or she controls. The foreman knows the people, the processes, the materials, and the machines and draws on this knowledge to solve the hour-to-hour and day-to-day problems encountered. It is natural, then, for the foreman to be concerned about having the necessary knowledge and experience when an industrial robot comes into his or her sphere of influence.

A large part of the foreman's job is handling hour-to-hour and day-to-day people problems. The successful foreman has developed a great deal of skill in interpersonal relationships to properly motivate, direct, and control those people supervised. The ability to perform this function defines the supervisor's worth.

While most foremen would gladly give up one or two real troublemakers within their organizations, any development which will remove people from their organizations naturally will be viewed as a threat. A product or advancement which totally removes people problems also removes one of the major reasons for the direct supervisor's job.

The rule of thumb with foremen and first line supervision is virtually the same as that with the hourly employees. The foreman must be brought into the program very early. The foreman's input must be considered carefully, so that he or she feels both knowledgeable and comfortable with the upcoming changes. Since the foreman is generally good at understanding and working with the hourly paid employees, his or her effort and guidance in properly presenting robots to subordinates can prove very valuable.

While sitting in an airport several months ago, I noticed that an older gentleman waiting for the same flight was looking through product literature on a line of industrial robots. Striking up a conversation with him, I found that he was returning from a two-day training session sponsored by one of the major robot manufacturers. He was a foreman of a small company that decided to investigate the use of robots. They chose him to visit the manufacturer and report back on his findings. This man was bubbling with excitement about the prospects of using robots in his organization.

It was obvious from the conversation that the few hundred dollars spent on the trip had generated such a level of enthusiasm that the expenditure would be repaid tenfold in increased efforts. The fact that his company would trust him with such an important assignment gave him a sense of pride and a determination to make robots work. Getting employees to participate actively in the evaluation, planning, and development of the robot installation will give you the greatest possible assurance of success.

The hourly employees associated with and working around the robot installation are the most knowledgeable people about the work being performed. The foreman or supervisor covering that area is the next best informed. While a successful installation of industrial robots without the hourly paid employees' cooperation is highly unlikely, I think it would be fair to say that a successful installation without the support of first line supervision is virtually impossible. The approach recommended here is not unique to installations of industrial robots but is simply good management technique in action. The main point is that good management technique is almost mandatory when implementing a change as new and different as an industrial robot.

Unions

If you have a union, you know that it is nothing more than production employees banding together, speaking with a single voice, and more than likely, carrying a big stick. Management within a unionized plant generally understands that production changes and improvements must be initiated with some concern for employee attitude. The union's basic job is to enhance and protect the jobs of its members. If the union views the installation of an industrial robot as endangering either the security or long-term growth potential of its members, it will resist the installation. If, on the other hand, the robot is viewed as an instrument to enhance security and provide new and better opportunities to union members, its adaptation and installation will be welcomed and encouraged.

At the Robots IV Conference held in Detroit from October 30 to November 1, 1979, the view of the UAW (United Automobile, Aerospace and Agricultural Implement Workers of America) was presented. This view is rather indicative of the approach we can expect from unions concerning industrial robots.

Unions today recognize the importance of increasing productivity. This can be seen in the following clause included in major contracts with the UAW:

> The improvement factor provided herein recognizes the principle that a continuing improvement in the standard of living of employees depends upon technological progress, better tools, methods, processes and equipment and a cooperative attitude on the part of all parties in such progress. It further recognizes the principle that to produce more with the same amount of human effort is a sound economic and social objective.

The union's ready acceptance of automation and increased productivity must not ignore the fact that job security is a very basic union goal. Security is the underlying theme of the union's policy on new technology, which was approved by the 1979 Collective Bargaining Convention attended by delegates from every UAW local. The union recognizes natural attrition as the generally accepted method of implementing changes that will reduce the overall size of the work force.

The union recognizes its own need to devise programs to protect workers against loss of skills and jobs. The UAW has maintained a policy that reducing the overall work week can provide a key element in creating job opportunities. A shorter work week without a loss in overall employee compensation can open new opportunities for maintaining jobs while providing individual workers with tangible benefits derived from the increased productivity.

Unions have not overlooked the importance of training programs to prepare their members properly for rapid technological changes. They have recognized the need to develop new job skills so that many of the skilled, technical, electronic, and computer jobs that will result from the implementation of new levels of computer controlled automation, including robotics, will fall within the domain of the bargaining unit. Most unions are demanding that these skilled jobs be manned by their members. Retraining of replaced workers to handle these new responsibilities is expected and demanded.

The unions believe that management must provide ample advanced notification of its intentions to automate to allow the union enough time to analyze and negotiate the intended changes properly. They believe that a full and open discussion of these plans must be conducted in an unhurried atmosphere so that they can fully assess the impact of the proposed changes on the bargaining unit.

As you can see, technological advancement seems to be accepted and in fact encouraged by the UAW, provided the current work force is assured of continuing job security as well as provided with new job opportunities as they arise. This view is likely to be the approach of most major unions and will provide a model that may be followed by many of the smaller organized labor groups.

Engineering and Top Management

Why would I put engineering and top management as a single consideration? The reason is a single, potential problem they have in common. That problem is the tendency to over-complicate and over-engineer robot installations. A large gap exists in industry between the opinions shared by the engineering and management group concerning what is feasible and reasonable for day-to-day production and the expectations of production management and production engineering people. Ideas which appear simple to the engineering oriented may appear impossible or impractical to production management and supervision. While developing applications for industrial robots, it is necessary to use a certain amount of common sense and prudence in determining the level of technology required. Make certain that you don't use a mini-computer when a 3" piece of hay bale wire can do the same job. Instead, develop a harmonious blend of the practicality of the production oriented personnel and the engineering expertise of the engineering department.

When approaching both top management and engineering personnel be sure to communicate your desire for a straightforward, simple, and reliable system. Resist their enthusiasm for very high technology solutions to low technology problems. While high technology robotic systems with sight, tactile sense, and so forth are very exciting for both the engineering discipline and the management people, they can easily become a nightmare for those who must maintain and operate these systems. The high technology applications have a lot of appeal, but the simple, basic solutions using robot technology provide the consistent return on investment and improved profitability desired. When dealing with these groups, it is important to identify and eliminate, as quickly as possible, proposals which involve technology for technology's sake. Instead, substitute imagination and innovation coupled with a driving desire for simplicity.

The approach to top management must involve realistic expectations. Resist the inclination to promise high levels of technical utilization in order to obtain project approval. If you don't, an engineering department will attempt to use the highest possible levels of technology to fulfill your unrealistic promises, setting the stage for disappointment and disillusionment. When approaching management with robot programs, eliminate the mysticism and magic associated with these machines. All of the fantasies that surround the name "robot" in science fiction cause management to attribute to these machines a higher level of capability than would be normally associated with other types of equipment. It is important to temper this inclination with a realistic view of actual performance. If you do not do this early in the implementation process, you'll travel a rougher than necessary road toward long-term success. So approach the installation from as practical and pragmatic a position as possible. Discuss the simplicity, practicality, and payback available from these machines. Make certain that management understands the operation and intended benefits of the robot installation. Understanding the installation and operation of a robot and the benefits possible is an important requirement for the continued support of top management. It is natural for management, especially nontechnical management, to expect more from industrial robots than they are capable of delivering.

In summary, the solution to this potential problem is exactly the same as the solution to any other human relations problems connected with industrial robots. That solution is a direct, accurate and complete communication of what is expected, what is intended, and what is required for the upcoming installation. In all areas of human relations, this early, honest, and complete disclosure of information provides a simple, yet important method of preventing potential problems.

Community

The families, friends, local suppliers, and local businesses in your community all have a strong interest in what happens within your company. Because of this, the potential exists for false information to spread about your use of industrial robots. This should not occur if you have taken the proper steps to inform the various management and supervisory groups as well as the hourly paid employees concerning the benefits, capabilities, and limitations of the installation. However, to prevent inaccurate and unwarranted bad publicity about robot installations, it is advisable to put some effort into communicating to the community a true image of your robotics efforts. This communication can take several forms.

Depending on the size of your company, news releases or press releases to the local press and television stations may be warranted. These releases should be worded carefully to indicate the benefits derived by both the employees and the company from the use of industrial robots. State a policy that no workers will lose their jobs because of the installation of industrial robots, and indicate that with industrial robots come enhanced opportunities for workers.

Another local community relations technique has proven successful for a number of companies. This technique involves holding an employee open house after the first robot is delivered but prior to its being placed into production. During this open house employees and their families can view the new machine in operation, with engineers or company officials available to answer questions. Employees and their families can see for themselves that the industrial robot is simply another piece of equipment with little or no magic about it. In this way, your company can lay to rest fear that a new, far advanced technology will threaten all job security within the firm.

If you have bulletin boards or employee newspapers or newsletters, announce in these forums, well in advance, that industrial robots will be employed. Again, the information should be complete and accurate, indicating the benefits to all parties to be derived from the use of industrial robots.

Placing product literature of the robot on company bulletin boards for employees to examine can be another useful technique. Be sure, however, to examine the literature carefully, since some manufacturers stress the advantages of robots over human labor to the extent that such literature might prove to be more harmful than helpful. Most robot manufacturers are well aware of the basic tenets of successful installations and realize that the support of all parties involved is necessary for success. Most have resisted appealing to the emotional desires of management to totally rid itself of all the burdensome labor problems. Nonetheless, some do stress the fact that robots do not take coffee breaks, get sick, slow down, or goof off. Product literature of this type should be carefully handled, and a certain concern for companies that market in this manner might be wise.

Again, the efforts required to properly communicate to hourly employees, unions, management, production supervision, and the community are a very small percentage of the overall effort necessary for a successful installation. This minimal effort, however, is a small price to pay to overcome the resistance to industrial robots that might very well develop should this effort be ignored.

Chapter Nine
COST JUSTIFICATION

Most companies, large and small, have some form of cost justification associated with the purchase of capital equipment. This justification procedure can be anything from a few handwritten notes to a very complete and detailed analysis. The purpose of all justification is the same.

Since few companies today have unlimited resources, they must make decisions on how available resources will be applied to various corporate needs. In most cases, this is done by developing a budget. The goings-on, prayer meetings, and corporate in-fighting that finally result in an annual budget are not important. Each company has its own formulas for arriving at the optimum balance. Gut-feeling hunches, management philosophy, competition, and the market all result in a series of decisions that dole out the available resources of a company. These decisions have a great deal to do with the success or failure of a firm.

One part of every budget is capital expenditure appropriations. This is the amount of funding available during the budget period for new capital expenditures. Once the amount of money available for capital equipment has been determined, it is simply a matter of looking at the number of requests for new equipment and comparing the two. If the amount of new equipment requested is less than the amount of money appropriated, everything balances nicely. I assure you that this has never happened. The wish list for equipment from each of the departments, divisions, plants, etc. always adds up to more than the amount budgeted, sometimes considerably more. Many different cost justification systems have been developed to decide which programs should be funded to maximize profitability. What do these programs have in common? Each attempts to determine the amount of return—in the form of increased profits, increased production, or decreased costs—per dollar of investment.

Any cost justification program must be divided into three separate categories. The first is a determination of the total investment requirement. The second measures the effect of the investment on operations, expenses, and profitability. The third is an analysis of the return in relation to the required investment. Figures 9.1 and 9.2 are forms which will assist in developing the total investment required to install an industrial robot. Figure 9.1 concerns itself with the direct cost of acquiring an industrial robot. A brief description of its various categories follows.

Robot Cost

Robot: This is the base cost or direct cost of the standard robot in question. It does not include any options, modifications, or changes. *Modifications*: This covers any special modifications to the machine performed by the robot vendor. The cost of special non-standard changes to the machine, changes in machine specifications or non-standard additions such as special computer interfaces, special software routines, and the like should be added here. *Options*: Most robots and, in fact, most machine tools have a long list of necessary options. "Necessary options" are those features which are not included in the base price of a machine, but which are required for its operation. Before you pick up stones to throw at the manufacturers, consider that in many cases it is simpler to provide a base machine which is only 80% complete and add the final 20% in the form of a variety of standard options which can be mixed or combined to provide the best possible machine for a particular customer's needs at the lowest possible cost. Be careful to determine the proper cost of an industrial robot and do not assume that the base price quoted by the manufacturer is the necessary configuration for your application. The various options required to perform the operations in question should be listed here with their associated costs.

Figure 9.1
Determine Investment Required

I. Robot Cost

Robot   _____

Modifications   _____

Options (List)

_____   _____

_____   _____

_____   _____

_____   _____

Training Costs   _____

Maintenance Supplies   _____

Test Equipment   _____

Total Robot Cost   _____

II. Tooling Costs

Hand or Gripper   _____

Fixtures   _____

Material Handling   _____

Total Tooling Costs   _____

III. Installation Costs

Mechanical   _____

Electrical   _____

Vendor Assistance   _____

Total Installation Costs   _____

IV. Engineering Costs

In-House Design
____Hrs. @ ____/Hr.   _____

Outside Engineering Design   _____

Programming
____Hrs. @ ____/Hr.   _____

Total Engineering Costs   _____

Total Investment Required ══════════════

Training Costs: If the cost of both program training and maintenance training is not included with the cost of the machinery, it should be added here. This training is a very necessary part of a robot installation. Its costs, not only the direct costs charged by the vendor, but also the travel and lodging expenses for those involved, should be considered as part of the investment in the robot installation.

Maintenance Supplies: Another expense which must be considered as part of the robot cost is the investment in various maintenance supplies and backup components. Unless down time of a day or more on the robot is acceptable, a stock of maintenance supplies and replacement parts will be necessary. Most robot manufacturers have a recommended spare parts list, and in general this should provide a good guideline as to what the investment in maintenance supplies will be.

Test Equipment: Certain installations may require various pieces of test or diagnostic equipment. While it is normally not necessary to purchase a piece of equipment as complex as an oscilloscope, especially if no one within your organization knows how to use it, items such as a digital voltmeter or continuity tester should be available to any robot installation. While the cost of most of these items is relatively small, it should be included as part of the cost of the robot installation.

Total Robot Costs: Add up all of the above items and place the total here. This total is the full cost of the robot equipment minus any special tooling directly related to this particular installation. These costs are more or less the fixed costs of a robot system of this type.

Tooling Costs

Hand or Gripper: A hand, gripper, manipulator, tool, torch, or the like will be necessary for the industrial robot to perform a useful task. The cost of any of this special tooling actually attached to the robot should be shown here.

Fixtures: In many applications some type of special fixturing separate from the robot itself may be necessary. These can be special holding fixtures in a spray painting application or special clamp fixtures in a welding application. Whenever these fixtures are necessary for an installation, their costs must be considered. Total the cost of any special fixturing required and place it here.

Material Handling Equipment: In many installations parts must be delivered to a point within the reach of the industrial robot in order for the installation to function properly. These expenditures include any types of conveyors, slides, part feeding devices, transfer devices, and the like. The cost of this related material handling equipment should be shown here.

Total Tooling Costs: Adding the last three lines will give you a value for the total tooling expenditure associated with this particular installation. This expenditure can vary greatly from installation to installation and is generally dependent on the complexity of the job being performed.

Installation Costs

Installation costs are all of those costs associated with the actual installation of the robot. They include the costs of site preparation, any special foundation work, utility drops and hookup, all of the interface devices connecting the robot to the various part feeders or conveyors, any rearrangement costs including relocation of various pieces of equipment and rerouting of stock, all changes to standard pieces of production equipment, and the like. The costs of necessary safety devices, fences, guardrails, and the like should also be included.

Mechanical: All direct and indirect mechanical labor, mechanical components, mechanical machine changes, and the like should be included here. Also include costs for relocation of various machines and fabrication of the necessary guardrails, safety fences, and the like.

Electrical: All electrical utility drops; power connections; interconnections between safety fences, various machines, and robot installation; and hookup of various limit switches and the like should be included here.

Vendor Assistance: While some robot vendors include the cost of a technician during startup as part of the initial machine costs, others do not. If there is a charge for having a vendor representative present during the installation startup, that cost should be considered here.

Total Installation Costs: Adding mechanical, electrical, and vendor assistance costs gives you the total installation cost. This total is the amount of effort and labor necessary to locate and hookup an industrial robot in a specific application.

Engineering Costs

Engineering efforts can be divided into two categories. The first of these is the design function. The design function determines precisely detail by detail how the robot will perform the necessary task. It includes deciding which wires will be connected together, which interconnects and safeties are required, and the like. This design effort can be conducted either in-house or by an outside engineering contractor.

The second form of engineering necessary for the implementation of a robotic system is the program development sequencing and machine startup.

In-House Design: The anticipated number of hours of in-house engineering effort required to develop and implement the robot application should be multiplied by the appropriate engineering cost per hour. This per hour cost should take into account the secondary cost of employment, including fringe benefits, unemployment insurance, taxes, and the appropriate indirect overhead expenses.

Outside Engineering Design: This area is reserved for those fees and associated expenses involved in hiring outside consultants or applications engineers to assist in the development and implementation of a robotics program.

Programming: The engineering or programming effort necessary to develop, debug, and back up the robot program properly should be indicated here. To develop programming costs, estimate the total number of engineering hours necessary and multiply that number by the programmer's cost per hour.

Total Engineering Costs: Find the total engineering cost by adding together the cost of in-house engineering design, outside engineering design, and programming.

Total Investment Required

You are now in a position to determine the total initial investment required to implement the robot application. Determine this figure by adding the total robot cost, the total tooling cost, the total installation cost, and the total engineering cost. This number then represents the funding necessary to implement the application in question.

One word of caution about determining the investment for the first robot application. In most instances the initial robot application will cost more than initially anticipated simply because of a lack of familiarity with the equipment, technologies, and procedures involved. As a safety factor, increase the investment requirement determined here by 20%–25% for the initial installation to allow for these unknown factors. This total investment figure will be used later in determining the overall efficiency of the installation in producing additional profits.

To complete your justification you must determine not only the savings associated with the installation but also the effect that the installation will have on the bottom line. It is tempting to simply look at the plus end of the equation, taking into consideration only the beneficial effects of a robot installation. In truth the cost associated with operating the installation must be subtracted from any benefits derived in order to obtain the actual effect of the installation in question. When trying to determine the economic benefits derived from a new technology installation, be certain to look at the entire picture and not at a narrow segment. Try to avoid shifting costs from one area or department to another and claiming the reduction in expenses of the first department as a true cost savings. This, unfortunately, is a common practice in large, multi-division companies in which a great deal of annual cost savings can be derived from simply restructuring and reorganizing the accounting system. You must try to develop an accurate and realistic assessment of measurable savings which should be derived from a successful installation. Figure 9.2 is a cost sheet titled "Determine Effect on Profits," which should help you assess savings accurately. Its categories are described below.

Operating Cost Per Year

Both the cost and savings from the installation will be calculated on an annual basis. If you wish, this cost and savings can be calculated on an hourly basis, daily basis, or on any other time frame you choose. Be certain, however, that all of the expenses and savings are calculated on the same basis.

Labor—Indirect: The total indirect labor necessary to support the daily and weekly operation of the robot should be indicated here. This would include labor that is necessary to load and unload chutes, provide material handling services to and from the robot, provide

Figure 9.2
Determine Effect on Profits

I. Operating Cost/Year

Labor	_____
Indirect	_____
Maintenance	_____
Programming	_____
Supplies	_____
Depreciation	_____
Robot	_____
Tooling	_____
Installation & Engineering	_____
Other Costs (List)	_____
	_____
	_____
	_____
Total Operating Costs	_____

II. Savings/Year

Direct Labor	_____
_____ Hrs. @ _____ /Hr.	
Indirect Labor	_____
_____ Hrs. @ _____ /Hr.	
Materials	_____
Quality Effects	_____
Reduced Rejects	_____
Reduced Rework	_____
Other Savings (List)	_____
	_____
	_____
Total Savings	_____

III. Production Capacity Effects

Sales Value	_____
Less	
Direct Materials	_____
Equals	_____
Times	
Increased Capacity	_____
Gives	
Production Capacity Effect	_____
Total Capacity Effect	_____
Total Return/Year	_____

cleanup of the area, provide tooling changes and adjustments, etc.

Labor—Maintenance: The estimated annual cost of maintenance and repair of the installation should be indicated here. This cost should be the number of maintenance hours expected, times the hourly costs of maintenance personnel including all auxiliary expenses, fringe benefits, and the like.

Labor—Programming: The estimated annual cost of developing and maintaining programs for the robot should be indicated here. This cost is the number of man hours expected each year, times the cost of a programmer (whether an engineer, production supervisor, or whatever), times the hourly cost of the individual in question including auxiliary expenses, fringe benefits, and the like.

Supplies: Supplies include any support materials, utilities, or services required each year to operate the proposed system.

Depreciation—Robot: The annual depreciation charge for the robot should be indicated here. It may be necessary to check with your accounting department to determine the method by which the depreciation is calculated, since there are several different types of depreciation used today. It will be necessary to determine what the depreciation cost will be on an annual basis. In those instances in which an accelerated depreciation schedule is used, it may be necessary to recalculate the depreciation each year, if you are looking at a multi-year payback. I will cover the long term effects of the investment more completely in the financial analysis section.

Depreciation—Tooling: Indicate here the depreciation per year for the fixtures and tooling as determined by your accounting department. This depreciation is separate from the robot depreciation, as many accounting departments use a different life or different form of depreciation for tooling than they do for direct capital equipment.

Depreciation—Installation and Engineering: If the installation and engineering expenses have been capitalized or will be capitalized by the accounting department, then the associated depreciation per year should appear here. If the installation and engineering expenses have been expensed or will be expensed, they will not show up as depreciation and need not be included here.

Other Costs: Include any additional or unique costs associated with the operation of the installation which have not been covered in any of the above categories.

Total Operating Costs: Adding all of the labor, supplies, depreciation, and other costs associated with the operation will yield the total operating costs. Keep in mind that these costs represent increases which would not be present if the robot installation were not made. In other words, you are not trying to take into account all of the operating and production expenses associated with the product, but only those expenses associated with the use of the robot installation.

Savings Per Year

Savings per year represents those savings or changes in the normal costs associated with production which are directly attributable to the installation of an industrial robot.

Direct Labor: The number of direct labor hours saved by the installation each year should be multiplied by the direct labor cost per hour, including auxiliary expenses, fringe benefits, and the like to develop the direct labor savings per year.

Indirect Labor: This area is reserved for any savings realized in indirect labor because of the robot installation. Such savings might include reduced clean-up requirements in spray painting applications, reduced tooling or die maintenance in some plastics operations, or the like. The estimated number of hours saved each year should be multiplied by the indirect labor costs, including auxiliary expenses and fringe benefits per hour, to obtain the indirect labor savings.

Materials: Any annual saving in direct production materials should be calculated and recorded here. Material savings might include reduced paint usage in a spray painting application, reduced die lubricant usage in a die casting operation, reductions in the amount of adhesive applied in an automatic gluing operation, or the like.

Quality Effects—Reduced Rejects: The increase in the number of parts which will not be rejected for quality reasons has a value which should be indicated here. The value of these parts is the sales value of the product, minus the direct material, times the number of parts saved from being rejected. The rationale behind this formula is explained in further detail in the "Production Capacity Effects" section.

Quality Effects—Reduced Rework: The reduced number of rejects resulting from automated production methods eliminates the need for materials and labor normally associated with rework. This material and labor savings should be calculated on the estimated reduction in rejects each year. Indicate the amount saved here.

Other Savings: This area is for any additional savings not covered above which will result from the installation. This includes areas such as investment tax credits; higher sales because of higher quality; reduced requirements for air makeup, exhaust, or emission controls in painting operations; reduced requirements for supplies such as gloves, safety shoes, safety glasses; etc.

Total Savings: Calculate the total savings from the installation on an annual basis by adding the direct labor savings, indirect labor savings, materials, quality effects, and other miscellaneous savings.

Production Capacity Effects

Assume that a certain amount of direct labor and overhead is expended each day. By dividing the total cost (labor and overhead) by the number of parts produced, you can determine the labor and overhead cost per part. If the number of parts produced increases while the labor and overhead stays the same, the labor and overhead associated with each part go down. However, calculating the change for each part produced, when the installation of an industrial robot increases capacity, can be a complicated and confusing affair. In order to simplify this, look at this situation from a different angle. Assume that a certain level of labor and overhead is associated with the production of 100 parts. If 120 parts are produced now with the same labor and overhead, the only additional cost for the production of the additional 20 parts is the raw material used. Since the labor and overhead of a single part times 100 parts pays for the entire labor and overhead expense, then it's reasonable to assume that the additional 20 parts resulting from increased efficiencies do not have labor and overhead costs associated with them. The profit made on those additional 20 parts is the sales price of the part, minus the cost. The cost in this case is only the raw materials involved.

In order to calculate the value of increased production capacity resulting from the installation of an industrial robot, take the sales value of a single part, subtract from that the direct materials associated with that part, and multiply the remainder by the number of additional parts which can be produced in a one year period as a result of the robot installation. The result of that multiplication is the production capacity effect.

Total Return on Investment

You have now determined both the investment required and the increased profits derived from the robot installation. Now it's time to analyze these figures. One simple and straightforward method is determining return on investment and payback period.

The payback period is the amount of time necessary for the savings generated by the new installation to pay for the installation. This payback period is calculated by dividing the total investment required by the total return per year. The result is the number of years necessary for the installation to pay for itself.

This calculation is not necessarily exact, since it does not take into account the effect of taxes. It therefore provides an idea of how long it would take the installation to pay for itself if the additional earnings or savings were not taxed. In order to determine the after-tax payback, use the same formula; however, reduce the total return per year by the amount of taxes which must be paid on that return at your present tax rate.

A return on investment which we refer to as the simple return on investment is the total return per year divided by the total investment required. The result of this calculation, as a percent, is referred to as the return on investment, or ROI.

Each of these simple calculations provides a good yardstick for determining which of a number of proposed capital expenditure projects should be funded. However, they make a number of assumptions which, depending on circumstances, may or may not be correct. These analyses assume that the entire investment required is spent immediately at the beginning of the program and that the operating costs and savings will remain constant year after year. The fact is that in today's volatile, inflationary world neither of these is exactly true. I would, therefore, like to suggest that an analysis based on the time value of money might be appropriate to identify the best use of available capital very carefully.

First let me explain the time value of money. This simple concept says that a dollar today and a dollar next year and a dollar five years from now each has a different value. A dollar today is obviously worth one dollar. A dollar next year, however, is not. A dollar today is actually worth more, when considered a year from now. For example, with an interest rate of 9% one dollar today is worth one dollar and nine cents a year from now. By the same token, one dollar and nine cents a year from now is worth one dollar today. So, future money is worth less than present money. This means that the point at which expenditures are made or savings are realized will have an overall effect on how great they really are when looked at in today's dollars. The easiest way to make a determination using the time value of money is to convert all monies, both present and future, into present day dollars. A dollar therefore, is worth a dollar; however, a dollar savings one year from now would be equivalent to only 94 cents at a 9% rate. A dollar savings a year from now is only worth 87 cents today at a 15% rate.

In order to assist you in making this evaluation, two charts are presented in the analysis form shown. The first chart, Figure 9.3, is the present value of future monies. The second chart, Figure 9.4, is the future value of present dollars. In order to develop a usable analysis based on the time value of money, you must complete the form in Figure 9.2 for each year of useful life of the project. Future profits and future investment should be increased to indicate the effects of inflation on each of the various areas. These numbers then should be converted to present value using the present value of future money chart provided. From this information the payback period and return on investment can be calculated again. Those installations in which either investment can be postponed or returns come the quickest will prove most valuable. While the efforts involved in this analysis are generally greater than most companies undertake, the results do an excellent job of zeroing in on those projects with the greatest real immediate potential for increasing the profits from available capital.

There are many other calculations, methods, and processes for justifying the expenditures necessary in a robot installation. In the end, however, someone must pass judgment. Seldom is this done strictly on raw mathematical data. The method used by many successful companies, both large and small—trusting "gut feeling" about whether or not a project will provide a benefit—may still be the most effective way of allocating capital.

Figure 9.3
Present Value of Future Dollars

Year	5%	6%	7%	8%	9%	10%	11%	12%	13%	14%	15%
1	1.00	1.00	1.00	1.00	1.00	1.00	1.00	1.00	1.00	1.00	1.00
2	.95	.94	.93	.93	.92	.91	.90	.89	.88	.88	.87
3	.91	.89	.87	.86	.84	.83	.81	.80	.78	.77	.76
4	.86	.84	.82	.79	.77	.75	.73	.71	.69	.67	.66
5	.82	.79	.76	.74	.71	.68	.66	.64	.61	.59	.57
6	.78	.75	.71	.68	.65	.62	.59	.57	.54	.52	.50
7	.75	.70	.67	.63	.60	.56	.53	.51	.48	.46	.43
8	.71	.67	.62	.58	.55	.51	.48	.45	.43	.40	.38
9	.68	.63	.58	.54	.50	.47	.43	.40	.38	.35	.33
10	.64	.59	.54	.50	.46	.42	.39	.36	.33	.31	.28

Figure 9.4
Future Value of Present Dollars

Year	5%	6%	7%	8%	9%	10%	11%	12%	13%	14%	15%
1	1.00	1.00	1.00	1.00	1.00	1.00	1.00	1.00	1.00	1.00	1.00
2	1.05	1.06	1.07	1.08	1.09	1.10	1.11	1.12	1.13	1.14	1.15
3	1.10	1.12	1.14	1.17	1.19	1.21	1.23	1.25	1.28	1.30	1.32
4	1.16	1.19	1.23	1.26	1.30	1.33	1.37	1.40	1.44	1.48	1.52
5	1.22	1.26	1.31	1.36	1.41	1.46	1.52	1.57	1.63	1.69	1.75
6	1.28	1.34	1.40	1.47	1.54	1.61	1.69	1.76	1.84	1.93	2.01
7	1.34	1.42	1.50	1.59	1.68	1.77	1.87	1.97	2.08	2.19	2.31
8	1.41	1.50	1.61	1.71	1.83	1.95	2.08	2.21	2.35	2.50	2.66
9	1.48	1.59	1.72	1.85	1.98	2.14	2.30	2.48	2.66	2.85	3.06
10	1.55	1.69	1.84	2.00	2.17	2.36	2.56	2.77	3.00	3.25	3.52

Chapter Ten
THE FUTURE

If you go back to 1970 you can find predictions by those associated with the robot industry which called for massive use of industrial robots by 1980. The need for robots, their demonstrated ability to perform a variety of industrial jobs, and their cost justifications seemed to indicate rapid growth right around the corner. After all, the computer industry exploded as soon as the technology was available, and it now was available for robots.

Twelve years later the results are dismal. The United States has an installed industrial robot base of less than 4,000. Robot manufacturers have been trying to develop the right formula, mixing price, technological capability, applications assistance, and a multitude of other factors needed to spur the promised growth.

Once again there are some very rosy projections for the robot industry. A number of authoritative sources are predicting compound growth rates of 35% per year. For the first time business publications, trade journals, and even local newspapers are writing about industrial robots. Both commercial television networks and the Public Broadcasting System have presented programming about industrial robots.

Is it for real this time? How do we know that the promise of industrial robots today won't result in the same disappointing performance that the promise of 1970 did? Is something different now? Is there a fundamental reason why industrial robots will be used in higher volumes through the 1980s?

To try to answer this question, I'd like to examine industrial robots as a competitive force in a free market system. Like most automation, this product competes with people. The industrial robot has some disadvantages from an economic standpoint when compared to hired employees. The major disadvantage is that an industrial robot is a piece of capital equipment. As such, it utilizes some of the limited resources discussed in the chapter on economic justification. People, on the other hand, can be hired and paid as necessary. They are an expense. This expense is directly proportional to the amount of production required. Should production requirements drop, people can be laid off, and the expense stops. The industrial robot in the same position continues to be an expense. Its depreciation continues, and there is no simple way to eliminate the cost. This inflexibility becomes even more pronounced when a large number of industrial robots are contemplated for a single operation. Because of this, as long as the necessary skills are available from people at a reasonable cost, the decision to utilize industrial robots can be postponed easily without adversely affecting production capabilities.

A Historical Perspective

To approach this question from another direction, let's go back to 1960 and examine people, specifically the number of people. Until 1960 the population of the United States grew each year at a steadily increasing rate. Because of government efforts and a multitude of other social and economic factors, 1960 became a turning point. While the population continued to grow after 1960, the rate at which it was growing leveled off and then began to decline.

Between 1970 and 1980 the work force expanded at an ever increasing rate. Those children born in the 1950s "baby boom" entered the work force. In addition, a sociological change occurred in the 1970s. Many more women entered, and remained in, the work force. Some of these were women interested in pursuing a long-term career. Others worked to counter the effects of inflation or to taste "the good life." The war in Southeast Asia ended during this time, and soldiers returned home to enter the work force. Throughout the

1970s the economic system faced the monumental task of finding work for this expanding labor force.

The government, which does react to the realities of life, even though it may not realize it, then stepped in. Additional bureaucracy, additional regulations such as pollution control, in fact a tremendous number of other programs reportedly motivated by the desire for a better world, slowed productivity growth, at times to a standstill. The fact is, large productivity increases through the '70s would have meant higher, possibly much higher, short term unemployment rates.

On a company-by-company basis, managers faced decisions on how to expand. The choice was whether to automate, reduce labor costs, and also bear the risks that come with adopting new technology, or whether to forego the potential profits of automation in favor of the current safe rate of production. How did American business make the choice?

In 1978 Herbert Simon, a psychology professor from Carnegie-Mellon University, won a Nobel prize. His study determined a rather interesting fact. When confronted with the choice between two or more alternatives, the American business executive does not, as people have been led to believe, choose the alternative which is most profitable. Faced with incomplete information the executive chooses the direction that is closest to a successful decision made in the past. The executive generally will forego profits or the promise of profits for the apparent security of knowing the outcome.

Because of this, American business in the '70s chose to use labor. In fact, in many instances labor was used instead of additional non-automated production equipment. This substitution of labor for capital seemed to make a lot of sense, since labor was abundant, not all that high-priced, and quite flexible and adaptable. With the growth in the gross national product throughout the '70s and the availability of labor to drive that growth, it is somewhat surprising that this country enjoyed any level of productivity growth at all.

In the environment of the 1970s, it is obvious that industrial robots could not fare well. A primary purpose of increasing productivity by decreasing labor content was not all that practical, when more than sufficient amounts of labor were available. This labor could be tapped, possibly at an ever increasing price, but without the technological risks or capital investment necessary for the use of industrial robots.

Future Population Patterns

What about the period between 1980 and 1990? It is twenty years since our population growth rate began to decrease. From 1980 until 1995 there will be fewer people entering the work force each year than the year before. In fact, in 1985 there will be a half million fewer young people entering the work force each year than in 1980, and by 1990 there will be a million fewer young people entering the work force each year than in 1980. Add to this the fact that many of the young couples who both worked in their '20s are now starting families in their '30s. While many of these women will continue to work after they have had their children, some will not. For those women who do continue to work, subtract working hours lost during pregnancy. All in all, the entire scenario reflects quite a reversal in the labor force.

In order to assess the impact of this on industry and on industrial robots, impose upon this slower growing work force the same general growth in gross national product. The only way for the scales to balance through the 1980s is by dramatic increases in the levels of productivity. One other way of balancing the equation would be to eliminate or drastically reduce growth in the gross national product for a ten year period. This has not ever been acceptable in the U.S., and I see no reason why the American public would settle for it now.

Just as the American government reacted to the realities of the population in the 1970s, it also must react to the changes in the 1980s. It must provide programs to enhance productivity. It must change from an anti-business to a pro-business status. It must begin to "reindustrialize" the country. To see these changes occurring, simply look at business publications and newspapers. Listen to the news on the radio and television. You will hear how the United States has slipped badly in comparison to Japan and Germany. You will hear how the country must provide incentives for business in order to increase productivity and "save" our American way of life. When the story is finally played out, most people will believe that the increased productivity and the resulting economic benefits were derived from some brilliant master plan developed by the federal government. The truth is, the increased productivity simply has to happen. The government simply is reacting to those forces beyond its control or comprehension.

Business will be no more eager to employ new technologies, take new chances, or move in new directions in the '80s than it was in the '70s. The difference, however, is that business will have less choice in the matter as time goes on. It is an economic fact that as the demand for labor increases, so will the cost of labor. Productivity increases will be necessary simply because there will not be a sufficient quantity of new labor at a reasonable cost to allow present production methods to expand without change. These effects will hit some industries worse than others and some sections of the country harder than others. However, the effect will be widespread, and few if any industries or companies will escape the effects of the change. I am aware of an instance during the last economic expansion in which a personnel agent from a furniture company in the Southeast actually went into neighborhoods knocking door to door attempting to find part time help for his factory. There are instances in the Southwest when machinery salespeople have been told that a customer would buy a new piece of equipment only if a salesperson could locate the labor to operate it. While the effect across the entire country will probably not be that severe, labor will indeed be more dear in the future.

The Factory of the Future

What effect will this phenomenon have on the use of industrial robots? Consider a typical small factory which has a variety of jobs. As in most factories, there are some relatively low-skilled jobs which are dirty, tedious, and relatively low paid. There also are some intermediate jobs requiring more decision-making power with slightly higher pay. There are jobs in an ascending order of complexity with ascending pay scales. Assuming a fixed amount of labor is available to run the factory, and assuming that this fixed amount of labor is not enough to fill all of the jobs, the least risky way to handle the situation is as follows. Since the higher-skilled jobs are those that produce the greatest amount of profit for the company, they need to be filled first. Should skills not be available within the firm, it might be necessary to train people presently doing a lower-skilled job to handle the higher-skilled job. This maximum utilization of available resources will ripple throughout the organization until the jobs left unfilled are the least desirable, lowest-paid, and least creative in the facility. Now it will be necessary to try to fill these positions from the outside. Filling them, however, will have become much more difficult, since most unskilled labor with any aptitude already has been trained by someone else for a higher level job. Those jobs for which training makes no difference are now relatively high-cost. Since the jobs are simple and repetitive, the solution that comes to mind is to automate the jobs. These will become the least risky jobs to automate from a technical standpoint. In this scenario many very simple tasks within industry will be turned over to robots, not because of a desire for profits but rather because of the need to solve a very basic problem, that is, getting the work out.

Now examine a particular job within the factory. Assume that the factory has a machine that manufactures widgets. The factory has been manufacturing widgets using the same machine and the same process for the last fifty years. Widgets are profitable and sales are up. The factory now needs to manufacture twice as many widgets. The obvious solution is to buy another widget-making machine, hire another operator, put the two together using well-established production methods, and produce twice as many widgets. As long as it is possible to hire the new widget operator, this is by far the surest solution to increased production and increased profits.

However, now impose the upcoming shortage of unskilled labor on this scenario. The unavailability of a widget operator will require management to automate the widget-making process somehow. In most cases this can be done in one of two ways. First, the factory could purchase a new, automated, super-duper widget-making machine which will produce twice as many widgets with one operator. The problem with this solution is that it changes the successful widget making process which has taken fifty years to perfect. The second possible solution is to find a mechanical substitute for the operator, that is, to come up with a robot that can do the simple loading and unloading tasks that an operator normally would do. This method has the advantage of maintaining all of the developed techniques and technology while solving the problem through automation. In the minds of most managers in industry today, the second solution looks much safer than the first. It is this scenario, multiplied hundreds of thousands of times within industry, that I believe will be one of the major forces behind the increasing use of industrial robots throughout the '80s.

The labor effect that I am describing will be very subtle, and a wave of robots will appear to be sweeping the country. While robots most certainly will create isolated instances of labor unrest, in general the problems anticipated with organized labor over the use of industrial robots will not develop. The reason for this is that robots will not be putting people out of work; instead they will be filling jobs which are unattractive or hard to fill.

Statements of this type obviously relate to the vast majority of applications. It always will be possible to find applications in which robots are doing jobs that people would rather have, and as the technology advances it will certainly be possible to observe robots performing tasks which require a certain level of judgment and decision-making capability. Staffing the vast majority of jobs, however, will reflect the tendency of decision-makers within industry to minimize risks. This fact will not change significantly through the years and will relegate the largest percentage of robots to the simple tasks for which unskilled labor is not available, economical, or necessary. The robots are coming, and this time they are being aided by the same forces that curtailed their use in the 1970s. The rosy predictions of the emerging robot market are, in fact, quite probable.

Future Trends in the Robot Industry

If the robots are coming, can we predict the way they will fit into industry? Can we predict how robots will be applied in five or ten years?

So far, this chapter has dealt with irreversible facts about the effects of population changes, interpreted from a little different viewpoint. However, as I begin to discuss how industry might apply industrial robots, I enter an area in which the views expressed must be completely subjective. The fact that industry must use robots in the coming decade is a fact that industry can do little to change. However, industry can influence the way in which these machines are used. Any prediction about the end result of literally millions upon millions of individual decisions that will determine the future certainly is risky. Despite this, I think it is possible to depict what logical decisions will bring us in the next few years, based on the past decision-making principles of American industry.

The present state of the robot industry consists of a multitude of development installations with some reasonable penetration in applications such as die casting. Today a high level of interest exists on the part of potential users. However, their lack of familiarity with these products causes concerns which hamper new installation attempts.

Robot manufacturers are providing a variety of different products and options in an attempt to develop the proper formula for success. Applications are still on a one-to-one basis, with the applications engineering work done separately for each installation.

The number of qualified robot applications engineers still is quite limited. Through the efforts of various manufacturers and industry organizations such as the Robot Institute of America and Robotics International, educational programs are now being offered. These programs, together with the large volume of literature being generated, should develop an awareness among engineers of the problems and solutions associated with industrial robots. Larger companies such as General Electric and General Motors sponsor in-house corporate robot seminars to instruct and educate their engineers in the proper application of industrial robots.

No general consensus has developed concerning these machines, although the market does seem to be segmenting itself by machine type and application. In the next few years industry efforts probably will continue to center on developing a common base of knowledge and information. The first step will be to develop a common language among various manufacturers, engineering disciplines, and end users. As I mentioned earlier, industry experts finally have formulated a definition of industrial robots. Despite disagreement over the accuracy of this definition, it certainly is a step in the right direction. In the near future more and more technical robotic terms will be defined officially.

As the need for industrial robots in simple applications begins to grow, the next step in this evolution probably will be the development of generic solutions to production problems. These packaged solutions will allow a particular industry to use robots in specific applications with little or no requirement for applications engineering. One example might be the development of a standard interface on a type of press which would allow virtually any industrial robot to be plugged into it. This combination of industrial robot and press then could operate as an automated system without a major applications engineering effort on the part of the end user. Any robot manufacturer that wanted to sell robots for this application then simply would build a machine that could be plugged into the press and do the job.

The key to these types of machine/robot combinations is the development of a standard operating method and a standard interface between various pieces of machinery being serviced by the robot and the industrial robot itself. Developing standard jobs and interfaces must be approached very cautiously. The tendency to standardize can stifle innovation within an industry. However, not developing these standards and interfaces once the necessary level of technology has been achieved can cost industry millions of wasted dollars in lost technological benefits.

Perhaps one way to begin this type of machine robot interface is to have the robots developed either by or in conjunction with manufacturers of machine tools or other systems with which the robot must operate. One possibility is a joint effort between a robot manufacturer and various machinery manufacturers and system houses. These arrangements can take a multitude of forms; however, each is able to multiply the applications engineering effort necessary to tie a robot into a system. Once the applications engineering has been developed, the machine and the robot labor to operate it are sold as an engineered package, thereby decreasing both cost and technical risk. A salesperson that runs into a customer who says, "Supply me with the labor, and I will buy your machine," then has a ready-made answer. At least one robot company has developed this type of arrangement, producing robots with other companies' names on them to be marketed as peripheral or optional equipment along with the company's standard product line.

I've been discussing the use of robots in the future concentrating on capacity-expanding applications. What about automating present production systems using industrial robots?

A large percentage of robot installations within the U.S. have been in situations in which some or most of the equipment with which the robot is to operate was already in place, although many times this equipment had to be relocated, rebuilt, or modified in some way. These types of applications obviously will continue to be important in the future. They also will continue to require extensive applications engineering work to make them function. As the various interfaces and machine robot systems are standardized and most pieces of new equipment are designed for robots, the advantages of a pre-engineered system will prevail. If the population of robots is going to have a 30%–35% growth rate per year, it will be impossible to spend an entire engineering design and development program on each robot installed.

Robots and Computers

To complete the scenario that I am projecting for sometime in the mid to late '80s, let me describe what is happening in the computer market and how the acceptance and proliferation of computers differs from the emerging market for industrial robots.

The technology that resulted in integrated circuits, tremendous miniaturization, and made today's computer market possible is a direct result of America's space program. The market for this technology grew so rapidly that industry soon was able to fund the huge research and development expenditures necessary for its continuing development. With new breakthroughs and increased volume came greater capabilities and lower costs. The price drop of typical electronic components used in computing was so dramatic that if automobiles followed the same price performance curve, a $35,000 Rolls Royce in 1965 would cost $2.30 in 1980 and get 280 miles per gallon. This is a truly remarkable feat, and everyone will agree that over the last ten years computers have changed all of our perceptions of life. As more and more computer capability became available, it was accepted almost immediately by every market approached, and dozens of major companies were born.

Why, then, has this not happened with industrial robots, whose capabilities have been available and underutilized for years? The answer to this question lies in the fact that computers have capabilities and functions not available from any other source. Without computers, it would be impossible to process the volume of information a typical company uses. The time it would take to gather, compile, and provide this information manually would make the information out of date by the time these tasks were completed. So in this case, computers provide something unavailable from a lower cost source. Robots, on the other hand, provide a service to industry in which in almost all cases people can replace, and people do not require a capital expenditure.

Once computers became accepted by a particular industry, the additional capability of these machines gave their users a competitive advantage over those not utilizing them. This provided another driving force behind the acceptance of computers. Competition dictated that when part of an industry began to utilize these new capabilities the remainder of the industry also was required to develop and implement these capabilities in order to remain competitive.

This scenario does not apply directly to industrial robots. The economic advantages of utilizing industrial robots do not provide additional competitive capabilities within an industry. Profits can be higher and labor costs and material costs can be lower, which might prove advantageous to those utilizing robots. However, until the cost of robots is reduced, the overall effects will not be a strong driving force behind their utilization.

The Technological Revolution

I have already indicated that through the '80s there will be pressure to increase the utilization of industrial robots based on shifts in population. As the mid and late '80s approach, it would be reasonable to expect that a large robot industry will emerge spurred by the rapidly growing market for their products. This market should be increased further by the fact that a proliferation of robot manufacturers and higher volume production levels will result in lower prices and better economics for those utilizing the robots.

Several other events also will occur simultaneously. There is no reason to expect that the rapid development of computer capabilities will slow. In fact, it is likely that computing power will continue to expand, and prices will continue to decline throughout this period. As this occurs, more and more capabilities can be built into inexpensive industrial robots. At the same time hundreds of industrial and academic laboratories working on various artificial intelligence systems will be developing some very real and unique capabilities for industrial robots. By the late 1980s the result of these efforts will be machines capable of performing various tasks with minimal instructions. An almost unbelievable level of "intelligence" will result. As this type of equipment becomes available for use in industry, a basic change in the robot market will occur. Through the '90s the robot market will change from the need-driven market of the early and mid-'80s to a desire-driven market much like today's computer market.

By the early '90s robots should be capable of providing many of a factory's skilled labor requirements with a minimal amount of applications engineering, minimal programming, and low cost. Utilizing people in these positions will seem totally illogical, yet this scenario does not foretell massive loss of jobs. There was a time not that many years ago when the largest segment of our population was employed in agriculture. Large numbers of people were required to provide the food needed. A revolution in farming, including chemistry for pesticides, fungicides, and fertilizers; machinery for preparation, planting, and harvesting; and improved packaging methods and distribution processes dramatically reduced employment in agriculture. This dramatic change in methods occurred in less than a generation, yet massive unemployment and tremendous suffering were not the result.

Massive unemployment and individual suffering will not be the result of the coming technological revolution, either. The 1990s will see the beginning of a change in the way the industrial world produces goods. The same dramatic reduction in employment experienced in the agricultural community at the turn of the century will be experienced at the end of this century and the beginning of the next. This massive change will bring with it the ability to produce huge amounts of low cost goods with little human effort. The efforts required will be in the form of transporting, storing, selling, and servicing these goods. As more capabilities result from the increased use of ever-expanding computer power, new industries will develop to use available human labor. If available workers and skills are unemployed, certain individuals within society will use the opportunity to organize these people into new industries. This is the role of the entrepreneur in society, and I assure you from personal experience these individuals are and will continue to be abundant. The physical changes I am predicting can be quite frightening when viewed as one large lump. The way they will occur, however, will not be frightening at all.

Major changes also will take place over a number of years in our way of thinking. As an illustration, consider some people's present attitude to farming. City dwellers may consider farmers as "hicks." However, most people agree that a farmer in an air-conditioned cab atop a one hundred thousand dollar tractor that is pulling a ten bottom plow would hardly qualify as a hick. The farmer is seen as a hard-working, resourceful entrepreneur. In fact, those who know farming would see the farmer as a very wealthy and self-sufficient individual. However, a farmer who cultivated fields using a mule and a plow certainly would not be taken seriously. Horse-drawn plowing and hand picking and harvesting are antiquated, inefficient, and unacceptable methods today. Someone attempting to use these methods to operate a farm would be regarded, and rightly so, as a fool.

By the turn of the next century this attitude toward the horse-drawn plow will be held by an increasing percentage of the population toward manual labor in factories. The labor necessary to produce goods slowly but surely will be regarded as fit only for machines, just as the labor necessary to plow the ground is regarded today as fit only for machines.

With the proliferation of interesting, challenging careers, the monotonous tasks of producing goods in industry will be relegated to machines. These machines will be produced using other machines. There still will be people within industry producing goods, just as there are still farmers growing food. These people, however, will be producing much larger quantities of products using their new technological tools. Their jobs will be more exciting, fulfilling, stimulating, and varied than the vast majority of production jobs in industry today. These changes will usher in an era of lower cost, high quality goods. The reprogrammable capabilities of industrial robots may even allow a certain level of customization of goods, since cycle-to-cycle variations can be built into production processes without sacrificing efficiency. This capability is not available with the hard automation techniques that are in use today for producing items in very large quantities.

I would not expect the relative cost of manufactured goods to be reduced much more than the relative cost of agricultural goods have been reduced over the last hundred years. However, this reduction in itself is quite significant. The incidental costs of obtaining manufactured goods, that is, the cost of packaging, transporting, and marketing these goods, will not be affected by these changes, just as incidental food processing, transportation, and marketing costs have not been directly affected by efficiency on the farm.

We can expect that this coming revolution in manufacturing methods will bring with it a higher standard of living. The benefits of this type of change will far outweigh the costs.

Yes, there will be some costs. Anytime there is a major shift of this type within a society, there is a transition period. This transition will be characterized by a retraining and re-education of a massive work force. It will require changes in the way the entire educational process operates within this country. Children must be trained to become young adults possessing the necessary skills to function in the new society. Our perception of the dignity of human effort will be altered, as we begin to prize it at a higher level than we do today.

Industry, not just in the U.S. but worldwide, is on the threshold of change. The changes that are coming can no more be stopped than could the industrial revolution. These changes will shape the future of humankind for the next generation and perhaps for the next several generations.

The future I have painted is certainly not the only possible direction for this industry, and of course I have not taken into account those major unexpected technical breakthroughs that make telling the future so difficult. However, I can guarantee that the industrial robot is now, and in the future will be, entrenched in world industry. The door has been opened, the creature is out, and we now must learn to live with it.

ROBOT APPLICATION
AND INSTALLATION MANUAL

This Robot Application and Installation Manual was developed to aid those companies that have embarked on a program to apply industrial robots in their production facilities. I have attempted to make this manual as universally applicable as possible and yet specific enough so that no major area of concern is left untouched.

The manual is designed as a workbook and is broken into a number of different sections. Each of these sections is comprised of a brief description of the section and a number of questions. To use the manual properly, you must bring together the various disciplines indicated in the introduction to each section and answer each question in writing in an open, free-flowing meeting. Many of the questions can be answered with a simple yes or no or with one or two words, but it is necessary to address each question individually.

If you employ this system properly, you should gain two valuable results. First, you should develop a thorough understanding of the application of industrial robots to your task so that hopefully you will be able to recognize, address, and solve potential problems during the planning stage. Second, the questions are structured in such a way that those participating in the meeting should develop an attitude toward implementing robots that has been highly successful in the past.

The development, implementation, and use of industrial robots is a new and very exciting field. These machines hold the promise of vastly improved productivity, higher quality, greater throughput, and ultimately higher profits. It is necessary, however, to understand that trying to implement the use of a robot without the proper foresight, thought, and planning can prove to be a disaster. I hope this manual can help prevent these disasters and aid in the planning and implementation of successful and profitable installations.

Defining the Task

The purpose of this section is to develop a thorough understanding of the task you wish the robot to perform. The following questions should be thoroughly and openly discussed in a meeting with Engineering, Production (on-line supervision, if possible), and Quality Control personnel. Developing written answers to the questions contained in this section will provide a necessary and valuable understanding of what is required of the industrial robot and will provide a firm foundation on which to build a successful installation.

1. Give a brief description of the task you wish the robot to perform.

2. In what general category does this task fall (material handling, spray painting, machining, etc.)?

3. What type of hand tooling will be required to perform this task?

4. How many outputs will be required to operate this hand tooling?

5. What weight(s) will the robot be required to carry in the performance of this task?

6. What is the rated capacity of the robot?

7. At what percentage of full capacity will the robot be operating?

8. Will the program required be point-to-point, continuous path, or a combination?

9. What method will be used to start the program?

10. Is the work environment sufficiently structured for the robot to operate properly?

11. What normal variations can occur?

12. How will these be handled?

13. What abnormal variations could occur?

14. What effect will these have on the system?

15. How many programs or program variations will be required?

16. How will these be sensed and elected?

17. Does your company have the technical expertise necessary for this level of programming?

18. How much time does it take a person to perform this task?

19. How much time will the robot have to perform this task?

20. How much of a safety factor is available?

21. What accuracy is required to perform the task?

22. What type of guide system could be used to allow for variation in robot positioning?

23. What happens if the robot malfunctions?

24. What happens if a machine other than a robot malfunctions?

25. Where will the robot physically be placed?

26. Can it reach all points necessary from that position?

27. Are there any pillars, columns, or other obstructions that must be considered?

28. Are there any machines, conveyers, or other equipment that must be relocated to accommodate the robot?

29. What benefits will result from the use of robots in this task?

30. What effect will the use of robots in this task have on the hourly personnel presently assigned to this task?

Developing the Plan

The questions in this section must be answered in order to develop and implement a plan to use industrial robots in the manufacturing process. Again a committee approach involving Engineering, Manufacturing, and Quality Control personnel provides the best method of answering these questions. I strongly recommend that the answers be written out and distributed to provide a common base of understanding of how the robot will be utilized.

1. When do you anticipate installing the robot?

2. What task will the robot be required to perform?

3. What type of hand configuration will be required?

4. Who is responsible for obtaining and following through with the installation of the hand?

5. Does the time required to secure the hand and tooling correspond with the desired date of installation?

6. List in detail the operating sequence of the robot and the machinery in the area in which it will operate. Indicate a time for each step in the robot program as well as for each operation occurring on the machinery which it services.

7. Who will be responsible for programming the robot?
   Why was this person chosen?
   Would it be desirable to have others in the organization learn how to program?

8. What system will be used to initiate the robot's cycle?
   Choose the program number.

9. What safety interlocks will be required?
   What signals must the robot receive from the equipment it services?
   What signals must the robot give to the equipment it services?
   How will these signal requirements be handled?

10. How much variation normally occurring in the work piece placement must the robot handle?
    How much variation does the hand configuration allow before it will malfunction?

11. Who will be responsible for the general installation of the robot?
    Who is responsible for providing electrical power?
    Who is responsible for installing the machine and bolting it in place?

12. What machinery must be moved or rearranged in order to accommodate the robot?
    What is the timetable for this work?
    Who is responsible for this job?

Personnel and Labor Relations

This section deals with the relationship of the robot installation to hourly paid labor, supervision, management, and the local community. This area is of substantial importance. Many robot installations have failed because of a lack of understanding and concern in this area. The following questions are designed to ensure that all concerned with the installation are informed in a timely and accurate manner. Again, the answers to these questions should be developed and written in a group meeting. Production personnel, including first line supervision in the area in which the robot will be placed, Engineering personnel, and those in Public Relations, Labor Relations, or Personnel all should attend this meeting. Make certain that every question is answered. Analyze the answers to make sure they make sense for the organization.

1. What total effect will the installation of a robot have on the number of hourly production workers?
   On their level of pay?
   On their job security?

2. What will happen to the individual directly replaced by the robot?

3. From the standpoint of the employees, how desirable would it be for them to perform the job(s) that will now be done by the robot?

4. How desirable are the new jobs to which they will be assigned?

5. Will any hourly employees be laid off 90 days prior to or following the installation of the robot?

   If so, will the installation of the robot get the blame for the lay-off?

   What effect will this have on the employees' morale and their desire to make the robot work?

6. When and how will the employees be informed about the installation of the robot?

   What mechanism has been established or is available to answer their questions about the robot installation?

7. What benefits do the hourly employees derive from the use of the robot?

   Have the employees been thoroughly and properly informed of these benefits?

8. Has first line supervision been involved with the robot and its program from inception?

   Are they thoroughly familiar and comfortable with the use of robots?

   Do they favor or oppose the use of robots in their areas?

9. Have all other first line supervision and middle managers been briefed on the robot installation, including its benefits, effects, etc.?

10. Do those responsible for programming the robot have the necessary level of technical experience?

    Have they been involved in the program on an on-going basis?

    Do they thoroughly understand how the installation will work, its purpose, and its benefits?

11. How will the local community be informed of your robot installation?

    If handled by word of mouth from your employees, will the information be accurate?

    If handled through news releases in local newspapers, who will write the releases and when will they be sent?

By what means will you ensure that the local community will learn of the benefits of the robot to your employees?

12. Is the robot installation getting the interest and scrutiny of top management?

    What effect will this have on those working on the program?

    Will an overly-cautious or overly-enthusiastic attitude toward the installation result?

13. Based on the method by which the installation is being handled, do the hourly employees other than those directly replaced by the robot receive the robot installation as an opportunity for advancement?

    As a threat to their job security?

    As a method to relieve some of their daily drudgery?

14. If this is a successful installation, what effects will it have from an employee relations standpoint on your ability to automate in the future?

15. If this is an unsuccessful installation, what effect will it have from an employee relations standpoint on your ability to automate in the future?

Installation

This section is designed to ensure that you have considered all the details necessary for a smooth installation and start up. Answers to the questions again should be written and developed in a meeting attended by Engineering, Production, and whichever organization will be responsible for the actual physical installation of the robot.

1. What will be the actual physical location of the robot?

   Of the hydraulic power unit?

   Of the control system?

2. Does this installation position meet all applicable codes (fire, safety, etc.)?

3. When do you expect to install the robot?

   Will production be required during this installation?

   If so, how will production be handled?

4. Who is responsible for the proper installation of the robot?
   Who is responsible for the start-up?
   Who is responsible for the on-going operation of the robot?
   Why were these particular people chosen?

5. Is the necessary power to operate the robot available at the installation site?
   If not, who is responsible for installing the service?
   When will this be done?

6. How will the robot be secured to the floor?
   Are all materials necessary to do this available?
   What effect will the normal problems and malfunctions during start-up have on plant production and productivity?
   Have the proper personnel, including top management, been alerted to this?

7. How will the electrical and hydraulic lines between the hydraulic power unit, actuator, and control cabinet be routed?

8. What length hydraulic or· electric lines is necessary for this installation?
   Are these supplied as standard equipment by the robot manufacturer?
   If not, have arrangements been made for extra length lines?

9. Based on the physical location of the machinery to be serviced, is there sufficient room for a person to program the robot by hand, or must it be programmed hydraulically?

Safety Considerations

Because of their almost human motions and actions, robots can instill a false sense of security in most employees. The robot is, however, a mindless piece of machinery which can malfunction and in doing so can cause serious injury to anyone within its reach. Therefore, it is imperative that all those who come in contact with the robot thoroughly understand all safety considerations. It is also vital that you give careful consideration to safety in the design and installation of industrial robots. Again, the questions in this section are designed so that most of the details of a safe installation will not be overlooked. These questions should be answered in writing during a meeting of Engineering, Production, and Safety personnel. If your organization has a safety committee with participants from the hourly work force, it is highly desirable that they be included in this meeting so that their input may be considered carefully.

1. Have all applicable electrical, fire, and safety codes been met?

2. What type of safety gate or interlock will be provided to keep personnel from wandering into the robot's sphere of operation?

3. If the robot should malfunction by not moving, what is the worst possible consequence?

4. If the robot should malfunction by moving in an unprogrammed position, what is the worst possible consequence?

5. Are the risks in the above two circumstances acceptable?
   If not, what can be done to allow for this contingency?

6. What safety interlocks are required between the machines serviced and the robot?
   What would happen if the interlocks failed?

7. Can the programmer program the robot without being exposed to unnecessary danger?
   If not, what could be done to make this a safer operation?

8. What effect would a momentary power failure have on the robot machine installation at the various stages of the cycle?
   If problems exist at certain points, could the installation be redesigned to eliminate these interferences?

9. Have first line supervision and hourly employees been briefed thoroughly on the hazards and safe operation of an industrial robot?

Backup and Contingency Plans

Although industrial robots historically have been highly reliable pieces of equipment, they can malfunction at times. The effect of such a malfunction can be minimal if proper backup and contingency plans have been made. In general, an industrial robot performs the function of a person. This means that, if necessary, a person can generally replace an industrial robot if it is not operational, therefore minimizing or eliminating the effect of the malfunction on the manufacturing facility. The time to make these contingency plans is when the robot is installed, prior to the first problem that arises. The questions in this section are designed to assist you in developing contingency plans. Again, they should be answered in writing during a meeting of Engineering, Production, and Personnel staff members.

1. What would be the effect on plant production if the robot did not function and no action was taken to correct it?

2. Can a person adequately perform the task that the robot is performing?
   What training or skills are required?
   What job classification must be used?

3. Are the necessary devices and interlocks installed so that the task can be performed safely by a human being?

4. Is there room for a person to perform the task without removing the robot?
   If not, how long would it take to remove the robot and provide physical access?
   Where would it be stored temporarily?
   What effect would this delay have on the operation of the production facility?

5. Are the tools, spray guns, etc., necessary for manual operation available, installed, and operating?

6. Where will the labor necessary to replace the robot come from?
   What effect will this have on productivity and plant operation?
   Have these people been informed of this possibility and trained to replace the robot, if necessary?

7. Who makes the decision to use the backup plan?
   Under what circumstances is this decision made?

8. Who is responsible for making sure that the robot is repaired?

9. What is the plan for putting the robot back into production when it is operational again?

Who is responsible for implementing this plan?

Economic Considerations

In the end, the decision to utilize a robot in industrial applications is generally made on economic grounds. To be selected, the robot generally must provide a more cost effective, profitable production method than any other. Therefore, to evaluate the economic impact of the installation of an industrial robot effectively, it is first necessary to understand economics as they presently exist, then determine the effect of introducing industrial robots. Once robots are installed, it is desirable to track the actual benefits derived from the installation. Approaching the economics of robots in this manner gives you a certain amount of control over the first installation and then provides detailed, accurate figures which provide the basis for justifying additional robots in the production facility. The questions in this section should be discussed in an open forum by members of the Engineering, Manufacturing, and Financial Planning groups within your company. The result of these discussions should be written answers to the questions that follow.

1. What is the task being evaluated?

2. Generally speaking, how will the robot perform this task?

3. Which robot have you selected to perform this task, and what is its base price?

4. Why have you selected that particular manufacturer and model of robot?

5. Are there any options, additions, or modifications to the base robot required so that the robot is capable of performing this task? If so, what will the cost of these additions or changes be?

6. What is the cost of using training and set-up personnel from the robot manufacturer?

7. What auxiliary equipment will be required from other manufacturers to enable the robot to perform this task?

Who are these manufacturers, and what are the prices of this auxiliary equipment?

Is this auxiliary equipment compatible with the robot selected?

8. What modifications are necessary to the existing work place or to existing equipment in order for the robot to perform this task? What is the cost of these modifications?

9. What are the estimated shipping costs of the robot and various auxiliary devices?

10. What is the approximate installation cost for the robot, including electrical, pneumatic, and hydraulic hookups, and in-house labor expense?

11. Can the installation be accomplished without interrupting production?

If not, what is the cost of the interrupted production?

12. What economic effect would start-up problems have on production levels?

What costs would result should such problems occur?

13. Who is providing the engineering for the installation, and what are the costs of these services?

14. What is the approximate total cost of acquiring and installing an industrial robot to perform this task?

15. What effect will the industrial robot have on the labor content of the task involved, and what savings or additional expense will result?

16. What are the material savings associated with the use of an industrial robot?

17. What are the incidental savings (reduced clean up, less air conditioning or ventilation, etc.) associated with the robot?

18. Is it reasonable to expect an increase or decrease in the level of product quality?

19. How much could the reject rate increase with the use of an industrial robot before all other savings are eliminated?

20. From an economic standpoint, what effect would the elimination of rejects at this operation have?

21. Will the production level for this task be increased by the use of an industrial robot?

22. Will an increased production level in this position affect the production levels in other parts of the operation?

23. What is the economic effect of this change in production level?

24. Based on the above, what is the approximate annual savings associated with the use of an industrial robot in this task?

25. Based on its cost and the annual savings, how long will it take for the robot to pay for itself?

GLOSSARY OF TERMS

The following terms are commonly used in automation and robotics today. I have attempted to provide a simple, understandable definition for each term. Since there are no accepted standards within the robot industry, the definitions presented here are the best, generalized definitions as accepted today by the largest number of informed individuals.

A

Accuracy
The difference between the point that a robot is trying to achieve and the actual resultant position. *Absolute accuracy* is the difference between a point instructed by the control system and the point actually achieved by the manipulator, while *repeat accuracy* is the cycle-to-cycle variation of the manipulator arm when aimed at the same point.

Actuator
An electrical, hydraulic, or pneumatic driver, such as a cylinder or electric motor, which delivers power for robot motion.

Adaptive Control
A method by which input from sensors automatically and continuously adjusts in an attempt to provide near optimum operation.

Analog-to-Digital Convertor (A/D)
An electronic device that senses a voltage signal and converts it to a corresponding digital signal (a string of 1s and 0s) for use by a digital computer system.

Android
A robot that approximates a human being in physical appearance.

169

Architecture
The basic composition and structure of a computer or other complex electro-mechanical device.

Arm
The joints, links, slides, and mechanical configuration of an industrial robot which supports a moving hand or tool.

Artificial Intelligence
The ability of a machine, computer, or mechanism to perform functions normally associated with human intelligence, such as decision-making, perception, problem solving, or pattern recognition.

Assembly Robot
A robot designed specifically for mating, fitting, or otherwise assembling various parts or components into either subassemblies or completed products. A class of generally small, lightweight, fast, accurate robots used primarily for grasping parts and mating or fitting them together.

Automation
The science and practice of machinery or mechanisms which are so self-controlled and automatic that manual input is not necessary during operation. The technique of making a process automatic or self-controlling.

B

Bang-Bang Robot
A non-servo controlled, point to point robot, which operates by "banging" into fixed stops in order to achieve the desired positions.

Any robot in which motions are controlled by driving each axis against a mechanical stop.

Base
The platform or structure of an industrial robot which is fixed and to which the various parts of the actuator are attached.

Batch Manufacturing
A process in which a facility produces a multitude of different parts by manufacturing them in groups, lots, or batches in which each part in the batch is identical.

Binary
A number system in which only two digits—1 and 0—are used.

Bit
In a digital system, the name of a single binary character, either a 1 or a 0. The smallest piece of data with which a digital computer can operate.

Byte
The name of a string of binary numbers generally eight bits long.

C

CAD (Computer Aided Design)
The use of a computer to develop the design of a product to be manufactured. The use of a computer to develop the design and necessary NC programs for use by the manufacturing equipment which will produce a product.

CAM
(Computer Aided Manufacturing)
The use of computers and computer technology to control,

manage, operate, and monitor manufacturing processes.

CID Camera
A solid state camera that uses a Charge Injection Imaging Device to transform the incoming image into digital form for computer use. The CID camera permits a readout of the image at any point any number of times.

CMOS (Complimentary Metal Oxide Semiconductor)
A family of electronic components known as integrated circuits which combine hundreds or thousands of components on a single, small electronic chip. This family of integrated circuits is characterized by moderate circuit density, moderate speed of operation, and very low power dissipation.

CNC (Computer Numerical Control)
The use of a computer to control a machine tool or piece of equipment. Programs may be developed, modified, or changed at the machine using the machine's control system.

CPU (Central Processing Unit)
That part of a computer which actually executes instructions and performs operations on the data.

Cartesian Coordinate System
A mathematical system for defining a point in space comprised of three linear perpendicular axes. These axes are normally called the X, Y, and Z axes. The X and Y axes define a horizontal plane, and the Z axis defines the vertical dimension.

Compliance
The ability of a mechanism to flex or comply, especially when subjected to external forces.

Contact Sensor
A device which sends a signal to the control system whenever mechanical contact is made with the device.

Controller
A device which controls an actuator by measuring the actual position, velocity, or other characteristics, then comparing this to a desired value and providing necessary control signals. That part of a machine tool or mechanism which provides the signals to operate the machine.

Cylindrical Coordinate Robot
A robot whose manipulator arm consists of a primary vertical slide axis on a rotary axis. The vertical slide axis and a second slide are at right angles to one another in such a way that the shape traced by a point at the end of the furthest axis at full extension is that of a cylinder.

D

D/A Convertor (Digital to Analog Convertor)
An electronic component which is capable of taking a digital signal from a controller or a computer and converting that to a corresponding analog voltage level.

DDC (Direct Digital Control)
Using a computer in a control loop to determine the difference

between the actual position and the desired position and convert that difference to a control signal in an attempt to adjust the motor, machine, or mechanism accordingly.

DNC (Direct Numerical Control)
The use of a central computer to store piece part programs and provide these to one or more remotely located NC machines via a communications link (wire).

Distributed Control
A control technique whereby the control for a single machine or process is located physically in two or more distinct places.

Dynamic Accuracy
The accuracy with which a control system can reproduce a condition (such as position) while it is in motion. The ability to reproduce motions accurately.

E

Encoder
A device normally located at the joints of a robot which converts the position of a joint or an axis to a distinct signal for processing by the robot's control computer or control system.

End Effector
A device connected to the end of a manipulator by which objects can be clamped, grabbed, or otherwise secured for movement.

F

Fixed Stop Robot
A robot whose axes are driven into mechanical stops in order to de-

velop their final positions. A nonservo, point to point industrial robot. (Also see *Bang-Bang Robot.*)

Flexible Manufacturing System
A manufacturing system whereby a group of machines, usually numerically controlled, is interconnected by a system of conveyors and part transport devices so that a variety of similar but different products can be manufactured automatically.

Force Sensor
A device, normally used on the wrist of the robot, which is capable of measuring a variety of different forces exerted and converting those forces to a signal for further processing by the computer control system.

G

Gripper
A clamp or device which an industrial robot uses to grasp or clamp an object.

Group Technology
A technique for grouping various parts into families based on their geometric shapes, processing sequence, or the like, so that these families may be processed together through mechanized or automated flexible manufacturing systems.

H

Hand
A clamp or gripper used to grasp objects that is attached to the end of the manipulator arm of an industrial robot.

Hardware
The actual, physical parts of an electronic control system, such as the printed circuit boards, electronic components, wiring, enclosures, etc. In contrast, computer programs themselves are called *software*.

Hierarchy
The division of elements in a control system based on their importance, with the higher levels taking priority and being processed before the lower levels.

I

Industrial Robot
A reprogrammable manipulator designed to perform functions in industry normally performed by a person, and to perform those functions in essentially the same manner and at approximately the same speed as a person. A reprogrammable, multi-function manipulator designed to move material, parts, tools or specialized devices through variable programmed motions in the performance of a variety of tasks.

Input/Output (I/O)
That part of a computer which provides the communication between the computer and other parts of the system or the outside world. A term denoting either input signals or output signals or both. Data involved in the communication between the system at-large and the computer or the outside world and the computer.

Integrated Circuit (IC)
A family of electronic circuits packaged in small units varying in size from less than 1/2" to over 2" square. Circuits can vary in complexity, from simple switching gates to very complex micro-processors. *Monolithic integrated circuits* are made with a single "chip" of semi-conductor material. A *hybrid integrated circuit* is made up of two or more such circuits connected in a single package.

Intelligent Robot
A category of robots that have sensory perception, making them capable of performing complex tasks which vary from cycle to cycle. Intelligent robots are capable of making decisions and modifications to each cycle.

Interactive
A type of computer control system in which an interaction occurs between the system operator and the system throughout processing. The computer asks various questions or provides choices to which the operator must react.

Interface
The name given to the method, procedure, or hardware which allows one system or part of a system to communicate with or work with another. As an example, the interface between a floor lamp and a house would be the electrical plug and electrical socket connecting the two.

J

Joint
That part of an industrial robot which moves the manipulator arm. That part of an industrial robot which provides a rotational or linear degree of freedom.

L

LSI
(**Large Scale Integration**)
A classification based on a level of complexity of an integrated circuit chip. Classifications range from *small scale integration (SSI)*, through *medium scale integration (MSI)* and *large scale integration (LSI)*, to *very large scale integration (VLSI)*.

Level of Automation
The degree to which a task or process operates automatically. This degree must take into account the ability of the system to diagnose problems in its operation, the ability of a system to recover from an error or fault, the ability of a system to start up and shut down without human intervention, and the like.

Limit Switch
An electrical switch with a trip arm or other device, positioned on a piece of equipment in such a manner that motion trips the switch and turns off power, thereby limiting the motion of the machine.

Limited Degree of Freedom Robot
A robot whose actuator contains fewer than six degrees of freedom.

Line Synchronization
The ability to synchronize the operation of an industrial robot with a moving production line so that the robot automatically compensates for variations in line speed.

Load Capacity
The maximum load which an industrial robot or machine can carry without a failure.

Long Term Repeatability
The degree to which an industrial robot or other programmable mechanism can repeatedly locate either the end points, the program path, or a cycle over a long period of time under the same conditions.

M

MIG (Metal Inert Gas) Welding
A method of joining two ferrous metal parts by passing a heavy electrical current from a metal rod to the grounded parts. The resulting electric discharge melts the metal rod and the part joints together to form a weld. This process normally is conducted with a shielding gas which prevents oxidation of the molten joint and thus increases weld integrity.

MOS
(**Metal Oxide Semiconductor**)
A type of semiconductor material used to manufacture certain integrated circuits.

Magnetic Core Memory
An older type of nonvolatile memory consisting of magnetic beads strung on current carrying conductors. A memory system which

uses magnetic polarization for storing and retrieving data.

Manipulation
Grasping, releasing, moving, transporting, or otherwise handling an object.

Manipulator
A mechanism usually with several degrees of freedom which is designed for grasping, releasing, moving, or transporting objects. The mechanical part of an industrial robot.

Master Slave Manipulator
A type of tellyoperator consisting of a master arm held, moved, and positioned by a person and a slave arm which duplicates the motions of the person at the same time the motions are being input. There is normally a scale factor between the master and slave arm so that the slave arm can be larger, reach farther, or carry more than the master arm.

Material Handling Robot
A robot designed to grasp, move, transport, or otherwise handle parts of materials in a manufacturing operation.

Material Processing Robot
A robot designed and programmed so that it can machine, cut, form, or in some way change the shape, function, or properties of the materials it handles between the time the materials are first grasped and the time they are released in a manufacturing process.

Mean Time Between Failure (MTBF)
The average time that a component device or piece of equipment can operate between failures.

Mean Time to Repair
The average time that a device, component, or machine is out of service each time a failure occurs.

Memory
Any device into which data can be input, retained, and later retrieved for use. That part of a computer which retains data or program information.

Microcomputer
A type of computer which utilizes a single chip micro-processor as its basic operating element. A system comprised of a micro-processor plus other necessary electronic elements to provide the input, processing, memory, and output necessary for computing.

Micro-processor
The basic element of a central processing unit developed on a single integrated circuit chip. A single integrated chip provides the basic core of a central processing unit, even though it may require additional components to operate as a central processing unit.

Minicomputer
A class of computer in which the basic element of the central processing unit is constructed of a number of discrete components and integrated circuits rather than being comprised of a single integrated circuit, as in the micro-processor.

Mobile Robot
A robot mounted on a movable platform. The motions of the robot about the work place are controlled by the robot's control system.

Multiplexer
A device which allows signals originating at a number of different places to be communicated to and processed through a single set of data lines by switching from one signal to the next sequentially.

Multiprocessor
A computer or control system comprised of two or more central processing units capable of simultaneously processing different programs or capable of dividing up and processing various tasks in a program.

Modular
Comprised of two or more standard sub parts which can be combined in a variety of ways to provide a variety of finished configurations. A construction method whereby the final device is comprised of a number of standard modules.

N

NC (Numerical Control)
A system for controlling machines in which the prerecorded information necessary for operating the machine is provided often through the use of paper tape or punch tape in coded form. The machine reads the numeric instructions and operates accordingly.

Nuclear Tellyoperator
A device used for manipulating objects or controls in a radioactive environment. This device normally consists of a master unit operated by a human operator and a remotely-located slave unit which duplicates the actions of the human operator.

O

Open Loop
A control system whereby data flows only from the control to the mechanism but does not flow from the mechanism back to the control.

P

PROM (Programmable Read Only Memory)
An electronic memory component into which data can be permanently stored. Once this data is entered, it then becomes nonvolatile and can be read any number of times. PROMs normally can be erased by exposing them to high levels of ultraviolet light for an extended period of time, at which point they can again be reprogrammed.

Pattern Recognition
The ability of a robot or computerized system to determine certain characteristics or structures within a picture. The ability of a machine to recognize patterns or shapes.

Payload
The highest possible weight that a machine or industrial robot can handle satisfactorily and continuously.

Performance
The degree to which a machine or mechanism is able to achieve a desired result.

Peripheral Equipment
Any piece of equipment separate from the central control or processing unit of a machine or a computer without which the system can still function, although possibly not fully. In computer systems this includes terminals, printers, magnetic tapes units, and the like.

Photoresistor
An electronic component which changes an electrical characteristic called resistance by an amount proportional to the amount of light striking the component.

Pick and Place Robot
A simple, point to point, non-servo, limited sequence robot designed primarily to manipulate objects from one place to another. (See also *Bang-Bang Robot* and *Fixed Stop Robot*.)

Piezo Electric
An electrical property of certain materials by which an electrical characteristic called impedance changes as the amount of mechanical pressure exerted on the material changes.

Pixel
An element of a picture. In order for a computer to analyze picture in sight systems, the picture is broken up into a series of picture elements called pixels. Each pixel is assigned a brightness level which is the average of the area in the pixel. In computer vision systems, the pixel then becomes the smallest area of resolution in a picture.

Playback Accuracy
The variation between the positions of a programmable system during initial teaching or programming and the resultant positions when the system attempts to playback the recorded information.

Polar Coordinate System
A mathematical coordinate system used to define positions on a plane using one linear and one circular axis. In this system a position is defined by the number of degrees of rotation from the zero position and the distance from the center of rotation.

Position Error
The difference between the position which a control system commands a servo mechanism to attain and the actual position attained.

Potentiometer
An encoder which provides a resistance proportional to position. This is achieved by running a brush over a resistant material and mechanically increasing or decreasing the amount of resistant material between the brush and the electrical connections.

Productivity
The measure of the amount of output, in either goods or services, per unit of input. The higher the productivity, the higher the output versus input.

Programmable
Capable of being instructed to operate in a specific manner by accepting commands from a remote source.

Programmable Controller
A control system often used to operate machinery in place of the standard electromechanical relays. The controls are programmed rather than permanently wired as in standard control methods.

Programmable Manipulator
A device capable of manipulating objects by executing commands stored in its memory.

Prosthetic Robot
A controlled mechanical device connected to the human body which provides a substitute for human arms or legs when their function is lost.

Proximity Sensor
An electronic device which senses when an object is within a specified short distance (normally a few inches or less) of the sensor.

R

RAM (Random Access Memory)
A volatile memory electronic component into which data can be stored and later retrieved, regardless of the location of the stored data.

ROM (Read Only Memory)
An electronic component used primarily as a memory storage device into which data is permanently stored during manufacture and from which this data may be read. Data held in ROMs is not

alterable during the normal operation of a system.

Rated Load
The amount of weight or mass for which a system was designed and with which a system or manipulator can operate with a specified margin of safety.

Real Time Control
A control system in which the calculations or control functions necessary for operation are conducted at the time that control is occurring, instead of pre-processing or predetermining the control responses.

Record Playback Robot
A robot programmed by recording the actual position of its joints along a desired trajectory. Running the program then plays these positions back through a servo control system.

Redundancy
Duplication of systems, devices, information, or hardware performing a specific task. Its primary objective increasing reliability, since any of the duplicated subsystems can accomplish the desired task.

Reliability
The ability of a device, system, or machine to operate properly over a period of time.

Remote Center Compliance (RCC)
A device fitted to the hand or grip of an industrial robot and designed primarily to provide compliance in assembling close fitting components, in which the compliance is about a point in space in

front of the part being assembled.

Repeatability
The ability of a system or mechanism to repeat the same motion or achieve the same points when presented with the same control signals. The cycle-to-cycle error of a system when trying to perform a specific task.

Resolution
The smallest increment of motion or distance that can be detected or controlled by the control system of a mechanism.

Resolver
An encoder system which converts rotary or linear mechanical positions to an analog electrical system. An encoder system capable of detecting very small changes in angular linear position.

S

SCR (Silicon Controlled Rectifier)
A solid state electronic control device by which small electrical currents can be utilized to control high electrical loads.

Semiconductor
A material in which the flow of electricity can be controlled, especially by external signals.

Sensory Controlled Robot
A robot whose control system is designed so that its motions are controlled by information sensed from the work place.

Sequence Robot
A robot whose physical motion and trajectory follows a pre-programmed sequence.

Servo Controlled
Controlled by a driving signal which is determined by the error between the mechanism's present position and the desired output position.

Servo Mechanism
A mechanical or electromechanical device whose driving signal is determined by the difference between the commanded position and the actual position at any point in time.

Servo Valve
A valve which produces a hydraulic fluid directly proportional to a low level signal input to the valve.

Shaft Encoder
A rotary encoder used to encode or determine the position of the rotary shaft.

Shoulder
The joint on an industrial robot that attaches the arm to the base.

Smart Sensor
The sensing device in which the output signal depends on preselected mathematical or logical operations which combine the sensory information with input from areas other than the sensor.

Software
A name given to instructions, programs, mathematical formulas, and the like utilized in the computer system, in contrast to the actual physical hardware of the system.

Solenoid
A device in which wire is coiled around a movable center core.

When current is passed through the wire, a magnetic field is established causing the core to be drawn in.

Solid State Camera
A television camera that uses a solid state integrated circuit to change the incoming light image into electronic signals.

Speed
The maximum speed at which the tip of a robot is capable of moving at full extension.

Spherical Coordinate Robot
A robot whose construction consists of a horizontally rotating base, a vertically rotating shoulder, and a linear transversing arm connected in such a way that the envelope traced by the end of the robot arm at full extension defines a sphere in space.

Spot Welding
A method of fastening sheet metal parts together in which a heavy electric current is passed through the plates at a spot. This current rapidly heats and melts the two sheet metal plates together, forming a small, round spot weld.

Stepping Motor
An electric motor which is designed to rotate in distinct steps with typically 200 steps per revolution. A stepping motor is driven by a digital pulse through a special driver circuit. In the absence of a driving pulse, a stepping motor is maintained in detent and will remain locked in that particular angular position.

Strain Gauge
A gauge normally made up of fine wires which when cemented to a component can measure very small amounts of motion caused by flexing. Strain gauges are utilized to measure strains and stresses in structural components. When cemented to an elastic material, strain gauges can act as force sensors.

T

TTL (Transistor Transistor Logic)
A signal processing system in which data in the form of low level electrical signals is processed through circuits either discretely or through integrated circuits comprised primarily of transistors.

Tachometer
A device capable of sensing the speed at which a shaft is rotating.

Tactile Sensor
A device, normally associated with the hand or gripper part of an industrial robot, which senses physical contact with an object, thus giving an industrial robot an artificial sense of touch.

Teleoperator
A master slave device which produces movements identical to or in direct proportion to actions or motions of a remotely located human operator. The device communicates certain feedback information, such as position, forces, etc. to the human operator.

Transducer
A device which converts one form of energy into another. Examples

include an electric pressure gauge which converts mechanical pressure into electrical signals and a photocell which converts light into electrical signals.

Transfer Machine
An apparatus, machine, or device designed primarily to grasp a work piece and move it from one stage of a manufacturing operation to another.

Transformations
A method of converting a point in one coordinate system to the same physical point in a different coordinate system. A mathematical conversion system used in providing industrial robots with line tracking ability.

V

Videcon
A trade name for a type of small vacuum tube commonly used in television cameras to change the incoming images into video signals.

Volatile Memory
A memory system in a computer or control system which requires a continual source of electric current to maintain the data it is storing intact. Removal of power from a volatile memory system results in the loss of the data being stored.

W

Wrist
That part of the manipulator of an industrial robot to which the hand or gripper is attached.